Insight on Multifractal Dynamics of ns-Laser Produced Plasmas

Insight on Multifractal Dynamics of ns-Laser Produced Plasmas

Stefan Andrei Irimiciuc

National Institute for Lasers, Plasma and Radiation Physics, Romania

Maricel Agop

Gheorghe Asachi Technical University of Iasi, Romania,
Romanian Scientists Academy, Romania

Ioan Merches

Alexandru Ioan Cuza University, Iasi, Romania

World Scientific

Published by

World Scientific Publishing Co. Pte. Ltd.

5 Toh Tuck Link, Singapore 596224

USA office: 27 Warren Street, Suite 401-402, Hackensack, NJ 07601

UK office: 57 Shelton Street, Covent Garden, London WC2H 9HE

Library of Congress Control Number: 2023941209

British Library Cataloguing-in-Publication Data
A catalogue record for this book is available from the British Library.

ISBN 978-981-127-066-6 (hardcover)
ISBN 978-981-127-067-3 (ebook for institutions)
ISBN 978-981-127-068-0 (ebook for individuals)

For any available supplementary material, please visit
https://www.worldscientific.com/worldscibooks/10.1142/13259#t=suppl

Desk Editor: Nur Syarfeena Binte Mohd Fauzi

Typeset by Stallion Press
Email: enquiries@stallionpress.com

Contents

Chapter 1

Laser Ablation: The Answer in Search of a Question

1.1. History of investigation techniques for transient plasma dynamics

Laser ablation is a series of processes that are the results of a laser beam impinging on a solid surface. The effects of "photonic ablation" (or the interaction of photons with matter) have been tentatively known for centuries. Some examples can be found even in Greek literature, such as the one recorded in 303 BC, in which the properties of a globe filled with water that can light fire, regardless of the environmental conditions, are presented. Also, the concept of "photonic ablation" is even mentioned by Archimedes, who proposed, in 203 BC, to reflect and focus the sunlight on the Phoenicians attacking the city using an array of mirrors (Miller, 1994).

The "modern" history of laser ablation started with a series of conference papers and talks during the 1960s. The results reported in those papers covered a series of fundamental aspects that are considered the pillars of laser ablation and laser-produced plasmas, which led to the development of an entirely new research direction. The first recorded "regular" paper was a theoretical study by Askar'yan and Moroz (1963), in which they made some calculations regarding the recoil pressure during the laser ablation of a solid target and discussed the acceleration of small particles or droplets in the framework of a "one-sided evaporation" model. They also predicted the presence of ultrasonic and hypersonic oscillations produced by modulated laser ablation. Later, in a similar pioneering experimental drive, Honig and Woolston (1963) reported some results from the investigation of laser ablation of various targets (metals, semiconductors and dielectrics). They reported, for the first time, a quantitative measurement of the ejected particles (3×10^{16} electrons and 10^8 positive ions per m^3). The published paper presented the first detailed study of the electron emission and its temporal profile. They analyzed the mass distribution using a modified commercial, double-focusing mass spectrometer, thus demonstrating the first use of the ion microprobe analysis. This study would subsequently become the basis for ion mass spectrometry and pave the way for electrical investigations of LPPs. In later papers, Lichtman and Ready (1963), using a simple assumption of thermionic emission, derived the temperature of a surface during laser–target interaction, finding values of about 3300 K for a ruby laser interaction with a tungsten target. Ready (1963) proved, for the first time, the implementation of high-speed photography as a viable method to study the temporal and spatial profiles of a plume of ejected material. The paper reported a carbon laser ablation plasma. One of the main results presented in that paper was that the emitted light from the plasma reached its maximum at about 120 ns after the start of the laser pulse and had an estimated lifetime of a few microseconds. From here, the expansion velocity of the plume was estimated as 20 km/s. Follow-up studies on carbon-based targets were performed by Howe (1963), who determined the energy of the ejected particles using the vibrational (0.86–1.72 eV) and rotational (0.38 eV) temperatures extracted from the fluorescence spectra of CN and C_2. This represents one of the first mentions of possible nonequilibrium conditions that were attributed to the cooling of the ejected particles during an adiabatic expansion. This subject was further investigated by Berkowitz and Chupka (1964), who observed, after post-ionization of the ablated plume, cluster ions of carbon ($n \sim 14$), boron ($n \sim 5$) and manganese ($n \sim 2$). Exploring the production of large structures during laser ablation, there have been reports (Neuman, 1964) of large "blobs of molten material" and "fragments of material" suggested by the first momentum transfer measurements. This short period of time is characterized by a fast expansion of laser–matter interaction and related topics, during which the first reported papers concentrated on the study of various properties of the ejected particles (electrons, ions, neutrals, clusters and emitted photons). This, coupled with the first estimations of plasma temperatures, velocities and densities, led to the formation of a coherent image of the complex processes involved in laser–matter interactions.

In the following years, there was a "boom" in articles focused on fundamental investigations of laser-produced plasmas performed over a wide range of laser characteristics (beam power, pulse width, repetition rate, etc.). The development of laser technology and measurement techniques led to more sophisticated experiments and more comprehensive theoretical models. Exciting results began to arise due to new means of study, such as visible, ultraviolet and X-ray emission measurements (Benavides *et al.*, 2016; Ehler and Weissler, 1966), coupled with the findings of multiply-charged ions (Archbold and Hughes, 1964) and two- and three-photon emissions (Sonnenberg *et al.*, 1964). All these achievements and findings led to the development of new applications that were proposed as alternatives to the existing ones.

In 1964, Berkowitz and Chupka (1964) proposed, for the first time, the laser ablation technique as an alternative to fusion, thus leading to the idea of laser confinement fusion. Another spectacular application that was born was pulsed laser deposition (PLD), as a response to the already existing sputtering techniques. Smith and Turner (1965) reported the first representative experiment with PLD. Although the authors experimented on a variety of materials using a ruby laser, the quality of the resulting thin films was secondary to that of the ones produced by sputtering. Not until the 1980s was laser film growth able to compete with the other well-established deposition techniques, when Dijkkamp *et al.* (1987) deposited a high-quality thin film of $YBa_2Cu_3O_7$. Since then, the PLD technique has been used to successfully produce thin films with a wide variety of properties, among which is a series of thin films with high crystallinity (ceramic oxides, nitrides, metallic multilayers) (Eason, 2007; Craciun *et al.*, 1994, 2005; Perriere *et al.*, 1994). The main advantages of PLD are the relatively low costs with respect to molecular beam epitaxy (MBE) or metal–organic chemical vapor deposition (MOCVD) and better control over stoichiometry and phase composition, which is very beneficial for the growth of complex materials, including high-quality nanomaterials that are impossible to synthesize otherwise. Some of the main successes of the technique can be summed up by the type of complex target that resulted (nanowires of Si and Ge (Morales, 1998), binary (In_2O_3; Li *et al.*, 2003), SnO_2 (Liu *et al.*, 2003), ZnO (Yang *et al.*, 2007) and ternary systems ($GaAs_{0.6}P_{0.4}$, $InAs_{0.5}P_{0.5}$, CdS_xSe_{1-x}, indium tin oxide (Savu and Joanni, 2006)) and more complex materials (Eisenhawer *et al.*, 2011).

As thin-film deposition technology flourished, the growing reliability and stability of commercial lasers, particularly Q-switched YAG lasers, improved the uniformity of film growth and the reproducibility of microprobe measurements. Significant progress was made, simultaneously, in the fundamental aspects of the deposition process. This was achieved through plume diagnostics and the development of theoretical models. The PLD process is a complex one. This complexity also comes from the correlations between a series of variables, including target composition, laser characteristics such as fluence, wavelength or pulse width, background gas species, the substrate's physical properties and overall PLD geometry. Changing one parameter often shifts the ideal settings for others. The effects of changing a single variable can be identified by keeping all other variables constant, and the variables are generally kept constant for simplicity. Due to this network of interrelationships, the control of the deposition process becomes complicated, as does the overall understanding of the LPP dynamics and how the properties of the plume can influence the final product. This image can somehow be simplified. Let us observe the deposition process from three different perspectives based on the possible influencing factors. One perspective covers the interactions between the laser beam and the target, governed by the physical properties of the target (reflectivity, thermal/electrical conductivity, heat of vaporization, etc.) and those of the laser beam (wavelength, pulse width, shape, etc.) (Benavides *et al.*, 2016; Zavestovskaya *et al.*, 2008). A second perspective describes the relationships between the physical properties of the target and the properties of the laser-produced plasmas (Hermann *et al.*, 2012; Williams *et al.*, 2008), and the third one describes the influence of the ejected particles on the properties of the resulting thin film. In order to gain some knowledge about any of these dependencies, it is imperative to use well-established investigation techniques (OES, ICCD fast camera imaging, Langmuir probe method, mass spectrometry, etc.) in order to find a unifying relationship between all these "variables".

The benefits of the proper use of the investigation techniques further led to the discovery of other spectacular results. The splitting of the plume is one of them, and it was first reported by Geohegan's group (Geohegan and Puretzky, 1995, 1996), when investigating the dynamics of LPP in an ambient gas. This group also proposed a theoretical description based on multiple-scattering and hydrodynamic approaches (Leboeuf *et al.*, 1996). The plume splitting has been further confirmed and studied by other groups (Harilal *et al.*, 2002, 2003; Wu *et al.*, 2013). All these results were obtained under typical PLD experimental conditions, i.e. a fluence of the order of 1 J/cm^2 and a background gas pressure of 1–100 mbar. We emphasize that similar results were also reported for laser ablation in vacuum (background pressure $< 10^{-5}$ mbar) and at fluences typically higher than 10 J/cm^2 (Amoruso *et al.*, 2010; Ursu *et al.*, 2009). From a theoretical perspective, the plume splitting is seen as the result of two distinct mechanisms for the particle ejection (Peterlongo *et al.*, 1994; Yoo *et al.*, 2000; Ursu *et al.*, 2009): The ions would be ejected on a very short time scale through a Coulomb process in the very intense field left by the electrons' laser excitation and detachment, while the neutrals would come from a subsequent thermal process (phase explosion; Kelly and Miotello, 1998), which needs more time to establish (Yoo *et al.*, 2000).

Besides the overall dynamics of the plume, looking closely at the individual dynamics of the ejected charged particles, a modulated behavior was observed. The first reports of were published in the 1980s. Borowitz *et al.* (1987) recorded a fast oscillation structure on the target current of about 100 ps period when irradiating with a 100 J nanosecond laser beam (fluence up to 10^5 J/cm^2). The first attempts at comprehending this "peculiar" behavior were based on the formation of single or multiple double layers in the close vicinity of the target. This picture was the main focus of a long series of papers reporting on charge separation in laser-produced plasma, mainly during 1970–1980 (Ludmirsky *et al.*, 1984, 1985; Pearlman and Dahlbacka, 1977). Eliezer and Hora (1989) gathered, in a very comprehensive manner, the state of the art of double and multiple layers in laser-produced plasmas. One of the remarkable results reported is an experimental proof with double-layer electric fields of 10^5–10^6 V/cm and widths of 10–100 Debye lengths (Eliezer and Ludmirsky, 1983).

The study of ionic and electronic multi-peak structure in laser-produced plasmas has had a resurgence in recent years when we can find reports on the presence of an multi-peak regime in the ionic current during the early stages of ablation ($< 1\mu$s). Our groups were at the forefront of this movement, having reported on the "peculiar" effects in LPP, double- and multiple-layer formations, as results of systematic experimental studies of plasma plumes generated by laser ablation in various temporal regimes (ns, ps and fs) in materials spanning from simple metals (Cu, Al, Mn, Ni, W, Te, In, Zn, Ti, etc.) to more complex compounds (ceramics, chalcogenide glasses, ferrites) (Gurlui *et al.*, 2006, Nica *et al.*, 2009, 2010; Ursu *et al.*, 2010; Pompilian *et al.*, 2014; Irimiciuc *et al.*, 2017; Focsa *et al.*, 2017). Besides these experimental studies, three theoretical approaches were proposed. One was based on the fractal model developed by Gurlui *et al.* (2008) as the interaction between two fractal structures and their corresponding interface (generally, this interface delineates the double layer) (Nica *et al.*, 2009), whereas the second one was based on differential physics (a collisional model based on the plasma ion frequency and electron–ion collision rate (Nica *et al.*, 2010) in the context of Lieberman's model for plasma immersion ion implantation (Lieberman, 1989)), and the final one was based on the AC Josephson effect (Gurlui *et al.*, 2008).

In this short introduction, we have attempted to review a few "firsts". All these results are considered building blocks, as the techniques implemented in the 1960s led to the development of new theoretical models and aspects of laser-produced plasmas never seen before. The history of laser ablation is full of "firsts": the first optical emission spectroscopy measurement led to the development of the laser-induced breakdown spectroscopy (LIBS) technique; the first measurements of the ionic energy distribution led to the further development of mass spectrometry; and the first picture of the ejected material foreshadowing the development of the ICCD fast camera imaging method. These are the pillars upon which the image that we have today of the laser ablation process as a whole is built. Now, we can see the effects of all these great moments in laser ablation history: The rise of fast camera photography paved the way for the plume splitting effects, while the probing of the charged particles led to the observation of plasma multi-peak structure.

As a way of moving forward, in this book, we attempted to use what history has taught us in order to achieve another "first". Optical emission spectroscopy and ICCD imaging coupled with Langmuir probe methods were implemented in a systematic way for the investigation of nanosecond, femtosecond and picosecond laser ablation plasmas generated in a series of simple (Al, Cu, Mn, Ti, Zn, Ni, In, Te and W) and complex $((GeSe_2)_{100-x} (Sb_2Se_3)_x)$ targets, with the aim of investigating the link between the properties of the target and those of the laser-produced plasmas. We found, for the first time, empirical relationships between the physical properties of the targets (atomic mass, electrical conductivity, heat of vaporization, melting temperature) and those of the laser-produced plasmas (expansion velocity, excitation temperature, ionic density, electron temperature, electron temperature) in all three ablation regimes. From a theoretical perspective, we expanded the unique fractal approach proposed by Gurlui *et al.* (2008) in order to simulate the spatial and temporal evolution of plasma parameters and derive, for the first time, a direct relationship between the fractal fluid and its specific parameters and the properties of the plasmas as they are recorded through investigation techniques.

References

Acquavivay S., Caricatoy A. P., De Giorgiy M. L., Dinescu G. A. L., and Perroney A. 1997. Evidence for CN in spectroscopic studies of laser-induced plasma during pulsed irradiation of graphite targets in nitrogen and ammonia Evidence for CN in spectroscopic studies of laser-induced plasma during pulsed irradiation of graphite targets in nitrogen. *J. Phys. B: At. Mol. Opt. Phys.*, 30, 4405–4414.

Agop M., Nica P. E., Gurlui S., Focsa C., Paun V. P., and Colotin M. 2009. Implications of an extended fractal hydrodynamic model. *Eur. Phys. J. D*, 56(3), 405–419.

Amoruso S., Schou J., Lunney J. G., and Phipps, C. 2010. *Ablation Plume Dynamics in a Background Gas.* pp. 665–676.

Archbold E. and Hughes T. P. 1964. Electron temperature in a laser-heated plasma. *Nature*, 204(4959), 670–670.

Askar'yan G. A. and Moroz E. M. 1963. Pressure on evaporation of matter in a radiation beam. *Sov. Phys.*, 13, 1638–1639.

Balika L., Focsa C., Gurlui S., Pellerin S., Pellerin N., Pagnon D., et al. 2013. Laser ablation in a running hall effect thruster for space propulsion. *Appl. Phys. A: Mater. Sci. Process.*, 112(1), 123–127.s

Bator M., Hu Y., Esposito M., Schneider C. W., Lippert T., and Wokaun A. 2012. Composition and species evolution in a laser-induced LuMnO3 plasma. *Appl. Surf. Sci.*, 258(23), 9355–9358.

Benavides O., de la Cruz May L., Mejia E. B., Ruz Hernandez J. A., and Flores Gil A. 2016. Laser wavelength effect on nanosecond laser light reflection in ablation of metals. *Laser Phys.*, 26(12), 126101.

Berkowitz J. and Chupka W. A. 1964. Mass spectrometric study of vapor ejected from graphite and other solids by focused laser beams. *J. Chem. Phys.*, 40(9), 2735–2736.

Bhattarai S. and Mishra L. N. 2017. Theoretical study of spherical langmuir probe in Maxwellian plasma. *Int. J. Phys.*, 5, 73–81. Science and Education Publishing; 2017 [cited 2017 Jun 7];5(3):73–81.

Borowitz J. L., Eliezer S., Gazit Y., Givon M., Jackel S., and Ludmirsky A. 1987. Temporally resolved target potential measurements in laser-target interactions. *J. Phys. D Appl. Phys.*, 14;20(2), 210–214.

Canulescu S., Papadopoulou E. L., Anglos D., Lippert T., Schneider C. W., and Wokaun A. 2009. Mechanisms of the laser plume expansion during the ablation of LiMn2O4. *J. Appl. Phys.*, 105(6), 063107.

Chen F. F. 1995. *Introduction to Plasma Physics.* Boston, MA: Springer US.

Chen F. F. 2003. *Lecture Notes on Langmuir Probe Diagnostics.*

Cremers D. A. and Radziemski L. J. 2006. *Handbook of Laser-Induced Breakdown Spectroscopy.* Oxford, UK: John Wiley & Sons Ltd.

Cristoforetti G., Tognoni E., and Gizzi L. A. 2013. Thermodynamic equilibrium states in laser-induced plasmas: From the general case to laser-induced break-down spectroscopy plasmas. *Spectrochim Acta — Part B: At. Spectrosc.*, 90, 1–22.

Dogar A. H., Ilyas B., Ullah S., Nadeem A., and Qayyum A. 2011. Langmuir probe measurements of Nd-YAG laser-produced copper plasmas. *IEEE Trans. Plasma Sci.*, 39, 897–900.

Doggett B. and Lunney J. G. 2009. Langmuir probe characterization of laser ablation plasmas. *J. Appl. Phys.*, 105(3):033306.

Donnelly T., Lunney J. G., Amoruso S., Bruzzese R., Wang X., and Ni X. 2010. Dynamics of the plumes produced by ultrafast laser ablation of metals. *J. Appl. Phys.*, 108(4), 0–13.

Donnelly T., Lunney J. G., Amoruso S., Bruzzese R., Wang X., and Phipps C. 2010. Plume dynamics in femtosecond laser ablation of metals. 643(2010), 643–655.

Doria D., Lorusso A., Belloni F., and Nassisi V. 2004. Characterization of a nonequilibrium XeCl laser-plasma by a movable Faraday cup. *Rev. Sci. Instrum.*, 75(2), 387.

Eason R. 2007. *Pulsed Laser Deposition of Thin Films: Applications-led Growth of Functional Materials.* Wiley-Interscience.

Ehler A. W. and Weissler G. L. 1966. Vacuum ultraviolet radiation from plasmas formed by a laser on metal surfaces. *Appl. Phys. Lett.*, 8(4), 89.

Eisenhawer B., Zhang D., Clavel R., Berger A., Michler J., and Christiansen S. 2011. Growth of doped silicon nanowires by pulsed laser deposition and their analysis by electron beam induced current imaging. *Nanotechnology*, 22(7), 075706.

Eliezer S. 1989. Double layers in laser-produced plasmas. *Phys Rep.*, 172(6), 339–407.

Eliezer S. and Ludmirsky A. 1983. Double layer (DL) formation in laser-produced plasma. *Laser Part Beams.*, 1(3), 251–269.

Focsa C., Nemec P., Ziskind M., Ursu C., Gurlui S., and Nazabal V. 2009. Laser ablation of AsxSe100−x chalcogenide glasses: Plume investigations. *Appl. Surf. Sci.*, 255(10), 5307–5311.

Focsa C., Gurlui S., Nica P., Agop M., and Ziskind M. 2017. Plume splitting and oscillatory behavior in transient plasmas generated by high-fluence laser ablation in vacuum. *Appl. Surf. Sci.*, 424(3), 299–309.

Fujimoto T. 1990. Validity criteria for local thermodynamic equilibrium in plasma spectroscopy. *Phys. Rev. A — At. Mol. Opt. Phys.*, 42(11), 6588–6601.

Geohegan D. B. and Puretzky A. A. 1995. Dynamics of laser ablation plume penetration through low pressure background gases. *Appl. Phys. Lett.*, 67(2), 197–9.

Geohegan D. B. and Puretzky A. A. 1996. Laser ablation plume thermalization dynamics in background gases: Combined imaging, optical absorption and emission spectroscopy, and ion probe measurements. *Appl. Surf. Sci.*, 96–98, 131–138.

Griem H. R. 2005. *Principles of Plasma Spectroscopy.* Cambridge Univ. Press.

Gurlui S., Sanduloviciu M., Strat M., Strat G., Mihesan C., and Ziskind M. 2006. Dynamic space charge structures in high fluence laser ablation plumes. *J. Optoelectron. Adv. Mater.*, 8(1), 148–51.

Gurlui S., Agop M., Nica P., Ziskind M., and Focsa C. 2008. Experimental and theoretical investigations of a laser-produced aluminum plasma. *Phys. Rev. E* 78(2), 026405.

Gurlui S., Sanduloviciu M., Mihesan C., Ziskind M., and Focsa C. 2009. Periodic phenomena in laser-ablation plasma plumes: A self-organization scenario. *AIP Conf. Proc.*, 821(1), 279–282.

Gurlui S. and Focsa C. 2011. Laser ablation transient plasma structures expansion in vacuum. *IEEE Trans. Plasma Sci.*, 39(11), 2820–2821.

Harilal S. S., Bindhu C. V., Tillack M. S., Najmabadi F., and Gaeris A. C. 2002. Plume splitting and sharpening in laser-produced aluminium plasma. *J. Phys. D Appl. Phys.*, 35(22), 2935–2938.

Harilal S. S., Bindhu C. V., Tillack M. S., Najmabadi F., and Gaeris A. C. 2003. Internal structure and expansion dynamics of laser ablation plumes into ambient gases. *J. Appl. Phys.*, 93(5), 2380–2388.

Harilal S. S., Farid N., Freeman J. R., Diwakar P. K., LaHaye N. L., and Hassanein A. 2014. Background gas collisional effects on expanding fs and ns laser ablation plumes. *Appl. Phys. A*, 117(1), 319–326.

Hansen T. N., Schou J., and Lunney J. G. 1999. Langmuir probe study of plasma expansion in pulsed laser ablation. *Appl. Phys. A*.

Hermann J., Mercadier L., Axente E., and Noël S. 2012. Properties of plasmas produced by short double pulse laser ablation of metals. *J. Phys. Conf. Ser.*, 399, 012006.

Honig R. E. and Woolston J. R. 1963. Laser-induced emission of electrons, ions, and neutral atoms from solid surfaces. *Appl. Phys. Lett.*, 2(7), 138.

Howe J. A. 1963. Observations on the maser-induced graphite jet. *J. Chem. Phys.*, 39(5), 1362–1363.

Irimiciuc S., Boidin R., Bulai G., Gurlui S., Nemec P., and Nazabal V. 2017. Laser ablation of (GeSe2)100-x (Sb2Se3)x chalcogenide glasses: Influence of the target composition on the plasma plume dynamics. *Appl. Surf. Sci.*, 418, 594–600.

Janesick J. R. 2013. Scientific charge-coupled devices. *J. Chem. Inform. Model*, 902.

Kelly R. and Dreyfus R. W. 1988. On the effect of Knudsen-layer formation on studies of vaporization, sputtering, and desorption. *Surf. Sci.*, 198(1–2), 263–276.

Kelly R. and Miotello A. 1998. On the role of thermal processes in sputtering and composition changes due to ions or laser pulses. *Nucl. Instrum. Methods Phys. B*, 141(1–4), 49–60.

Kramida A., Ralchenko Y., and Reader J. 2014. *NIST Atomic Spectra Database Lines Form*. NIST ASD Team.

Koopman D. W. 1971. Langmuir probe and microwave measurements of the properties of streaming plasmas generated by focused laser pulses. *Phys. Fluids*, 14(8), 1707.

Lahm S. H. 1965. Unified theory of langmuir probe. *Phys. Fluids*, 8, 73.

Leboeuf J. N., Chen K. R., Donato J. M., Geohegan D. B., Liu C. L., and Puretzky A. A. 1996. Modeling of dynamical processes in laser ablation. *Appl. Surf. Sci.*, 96–98, 14–23.

Lichtman D and Ready J. F. 1963. Laser beam induced electron emission. *Phys. Rev. Lett.*, 10(8), 342–345.

Liu X., Du D., and Mourou G. 1997. Laser ablation and micromachining with ultrashort laser pulses. *IEEE J. Quantum Electron.*, 33(10), 1706–1716.

Liu Z., Zhang D., Han S., Li C., Tang T., and Jin W. 2003. Laser ablation synthesis and electron transport studies of tin oxide nanowires. *Adv. Mater.*, 15(20), 1754–1757.

Lochte-Holtgreven W., North-Holland E., Ovsyannikov A. A., and Zhukov M. F. 2000. *Plasma Diagnostics*. Cambridge International Science Publishing.

Ludmirsky A., Givon M., Eliezer S., Gazit Y., Jackel S., and Krumbein A. 1984. Electro-optical measurements of high potentials in laser produced plasmas with fast time resolution. *Laser Part Beams.*, 2(2), 245–250.

Ludmirsky A., Eliezer S., Arad B., Borowitz A., Gazit Y., and Jackel S. 1985. Experimental evidence of charge separation (double layer) in laser-produced plasmas. *IEEE Trans. Plasma Sci.*, 13(3), 132–134.

Merlino R. L. 2007. Understanding Langmuir probe current-voltage characteristics. *Am. J. Phys.*, 75(12), 1078.

Mihesan C., Lebrun N., Ziskind M., Chazallon B., Focsa C., and Destombes J. L. 2004. IR laser resonant desorption of formaldehyde-H2O ices: Hydrated cluster formation and velocity distribution. *Surf. Sci.*, 566–568, 650–658.

Mihaila I, Ursu C., Gegiuc A., and Popa G. 2010. Diagnostics of plasma plume produced by laser ablation using ICCD imaging and transient electrical probe technique. *J. Phys. Conf. Ser.*, 207, 012005.

Miller C. J. 1994. *Laser Ablation*. Berlin, Heidelberg: Springer-Verlag.

Morales L. 1998. A laser ablation method for the synthesis of crystalline semiconductor nanowires. *Science*, 279(5348), 208–211.

Mott-Smith H. M. and Langmuir I. 1926. The theory of collectors in gaseous discharges. *Phys. Rev.*, 28(4), 727–763.

Neuman F. 1964. Momentum transfer and cratering effects produced by giant laser pulses. *Appl. Phys. Lett.*, 4(9), 167–169.

Nica P., Vizureanu P., Agop M., Gurlui S., Focsa C., and Forna N. 2009. Experimental and theoretical aspects of aluminum expanding laser plasma. *Jpn. J. Appl. Phys.*, 48, 1–7.

Nica P., Agop M., Gurlui S., and Focsa C. 2010. Oscillatory langmuir probe ion current in laser-produced plasma expansion. *EPL (Europhysics Lett.)*, 89(6), 65001.

Ovsyannikov A. A. and Zhukov M. F. 2000. *Plasma Diagnostics*. Cambridge International Science Publishing.

Pearlman J. S. and Dahlbacka G. H. 1977. Charge separation and target voltages in laser-produced plasmas. *Appl. Phys. Lett.*, 31(7), 414–417.

Pompilian O. G., Dascalu G., Mihaila I., Gurlui S., Olivier M., and Nemec P. 2014. Pulsed laser deposition of rare-earth-doped gallium lanthanum sulphide chalcogenide glass thin films. *Appl. Phys. A: Mater. Sci. Process.*, 117(1).

Pompilian O. G., Gurlui S., Nemec P., Nazabal V., Ziskind M., and Focsa C. 2013. Plasma diagnostics in pulsed laser deposition of GaLaS chalcogenides. *Appl. Surf. Sci.*, 278, 352–356.

Puretzky A. A., Geohegan D. B., Fan X., and Pennycook S. J. 2000. Dynamics of single-wall carbon nanotube synthesis by laser vaporization. *Appl. Phys. A: Mater. Sci. Process.*, 70(2), 153–160.

Puretzky A. A., Geohegan D. B., Fan X., and Pennycook S. J. 2000. In situ imaging and spectroscopy of single-wall carbon nanotube synthesis by laser vaporization. *Appl. Phys. Lett.*, 76(2), 182–184.

Puretzky A. A., Geohegan D. B., Haufler R. E., Hettich R. L., Zheng X. Y., and Compton R. N. 1993. Laser ablation of graphite in different buffer gases. *AIP Conf. Proc.*, 288(1993), 365–374.

Ready J. F. 1963. Development of plume of material vaporized by giant-pulse laser. *Appl. Phys. Lett.*, (1), 11–13.

Savu R. and Joanni E. 2006. Low-temperature, self-nucleated growth of indium–tin oxide nanostructures by pulsed laser deposition on amorphous substrates. *Scripta Mater.*, 55(11), 979–981.

Schou J. 2009. Physical aspects of the pulsed laser deposition technique: The stoichiometric transfer of material from target to film. *Appl. Surf. Sci.*, 255(10), 5191–5198.

Smith H. M. and Turner A. F. 1965. Vacuum deposited thin films using a ruby laser. *Appl. Opt.*, 4(1), 147.

Sonnenberg H., Heffner H., and Spicer W. 1964. Two-photon photoelectric effect in Cs3Sb1. *Appl. Phys. Lett.*, 5(5), 95–96.

Sunil S., Kumar A., Singh R. K., and Subramanian K. P. 2008. Measurements of electron temperature and density of multi-component plasma plume formed by laser-blow-off of LiF-C film. *J. Phys. D: Appl. Phys.*, 41(8), 085211.

Tankosić D., Popović L. C., and Dimitrijević M. S. 2001. Electron-impact stark broadening parameters for ti ii and ti iii spectral lines. *At. Data Nucl. Data Tables*, 77(2), 277–310.

Tang E., Xiang S., Yang M., and Li L. 2012. Sweep langmuir probe and triple probe diagnostics for transient plasma produced by hypervelocity impact. *Plasma Sci. Technol.*, 14(8), 747–753.

Ursu C., Gurlui S., Focsa C., and Popa G. 2009. Space- and time-resolved optical diagnosis for the study of laser ablation plasma dynamics. *Nucl. Instrum. Methods Phys. Res. Sect. B: Beam Interact. Mater. At.*, 267(2), 446–450.

Ursu C., Pompilian O. G., Gurlui S., Nica P., Agop M., and Dudeck M. 2010. Al2O3 ceramics under high-fluence irradiation: plasma plume dynamics through space- and time-resolved optical emission spectroscopy. *Appl. Phys A*, 101(1), 153–159.

Ursu C. 2010. *Caracterisation par methodes optiques et electriques du plasma produit par ablation laser.* Université Lille 1 — Sciences Et Technologies.

Williams G. O., O'Connor G. M., Mannion P. T., and Glynn T. J. 2008. Langmuir probe investigation of surface contamination effects on metals during femtosecond laser ablation. *Appl. Surf. Sci.*, 254, 5921–5926.

Wu J., Li X., Wei W., Jia S., and Qiu A. 2013. Understanding plume splitting of laser ablated plasma: A view from ion distribution dynamics. *Phys. Plasmas*, 20(11), 113512.

Yao Y. L., Chen H., and Zhang W. 2005. Time scale effects in laser material removal: A review. *Int. J. Adv. Manuf. Technol.*, 26(5), 598–608.

Yang R., Chueh Y.-L., Ruth Morber J., Snyder R., Chou L.-J., and Wang Z. L. 2007. Single-crystalline branched zinc phosphide nanostructures: Synthesis, properties, and optoelectronic devices. *Nano Lett.*, 7(2), 269–275.

Yoo J. H., Jeong S. H., Mao X. L., Greif R., and Russo R. E. 2000. Evidence for phase-explosion and generation of large particles during high power nanosecond laser ablation of silicon. *Appl. Phys. Lett.*, 76(6), 783.

Zavestovskaya I. N., Glazov O. A., and Demchenko N. N. 2008. Threshold characteristics of ultrashort laser pulse ablation of metals. *Proceedings of the 3rd International Conference Frontiers of Plasma Physics and Technology*, pp. 10–18.

Zimmermann F. M. and Ho W. 1995. State resolved studies of photochemical dynamics at surfaces. *Surf. Sci. Rep.*, 22(4–6), 127–247.

Chapter 2

Diagnostics Tools for Transient Plasmas Generated by Laser Ablation

In the previous chapter, we presented briefly the fundamental processes that take place when a pulsed laser beam interacts with a solid target. The complexity of those phenomena and the strong dependence on the beam properties (wavelength, pulse width, repetition rate) (Yao *et al.*, 2005; Hussein *et al.*, 2013; Craciun *et al.*, 1995) and external conditions (background pressure, target bias and target physical properties) (Liu *et al.*, 1997; Amoruso *et al.*, 2010; Donnelly *et al.*, 2010; Harilal *et al.*, 2014) make it somehow difficult to have a profound understanding of the laser beam–target–plasma relationships, which transcends one set of experimental conditions and relates to all ablation regimes and different types of targets. In order to shed some light on either of these relations (laser–target, target–plasma or laser–plasma), over the time, various investigation techniques were implemented and adapted for the study of laser-produced plasmas (Geohegan *et al.*, 1992; Harilal *et al.*, 2003; Doggett *et al.*, 2009; Dzierga *et al.*, 2006, 2010).

Regarding the diagnosis of laser-produced plasma, the aim is to gather knowledge about the chemical composition of the plasma plume (atoms, molecules, nanoparticles, clusters, etc.), the electrical composition (ionization degree of the ejection particles, electron number density, etc.), the plasma plume dynamics (overall plasma dynamics or the dynamics of the individual plasma species) and plasma energy (in terms of electronic, excitation, vibrational or rotational temperatures). We emphasize that, due to the particularities of laser-produced plasmas, there is a requirement for all these plasma parameters to be space- and time-resolved. The current available plasma diagnostic methods can be divided into two categories: optical (fast camera photography (Harilal *et al.*, 2003), interferometry (Donnelly *et al.*, 2010), shadowgraphy (Feinaeugle *et al.*, 2012), optical emission spectroscopy (Geohegan *et al.*, 1992; Briand *et al.*, 2011), laser-induced fluorescence (Sappey *et al.*, 1991), Thomson scattering (Dzierga *et al.*, 2006, 2010; Travaillé *et al.*, 2011)) and electrical methods (mass spectrometry (Chen *et al.*, 2014), electrostatic analyzers (Ursu and Nica, 2013), Langmuir probes (Doggett *et al.*, 2009), and Faraday cups (Doria *et al.*, 2004)). The task of obtaining a complete description of the laser-produced plasmas is difficult/laborious due to their transient character. The typical lifetime of the plume is on the tens of microsecond scale, although faster phenomena can occur even in the sub-nanosecond timescale (Eliezer and Hora, 1989; Gurlui *et al.*, 2008). Accordingly, the investigation methods have to be selected carefully and adjusted to the "requirements" of the experimental configuration, depending on the analyzed phenomena (e.g. the Langmuir probe method is usually adapted for the space–time fast expansion of the plume in what is known in literature as the "sweeping" approach (Koopman, 1971), which will be presented in detail in Section 2.2.3). In addition, one has to take into account the intrinsic limitations of each technique, as all of them are usually not applicable throughout the whole spatial and temporal evolution of the plasma plume. For example, the Thomson scattering technique requires higher electron densities than the ones required by the Langmuir probe technique. In Figure 2.1, we represented the coverage, in terms of space–time coordinates, of some of the mentioned techniques. It can easily be seen that there is no "perfect" or "best" technique, but only the right method for the purpose of the experiment.

    Each technique reveals a unique facet of the plasma dynamics and provides important information that can be translated into diverse applications. For the deposition of thin films by laser ablation, parameters like angular distribution, particle density, particle velocity or spatial distribution are relevant to the properties and the

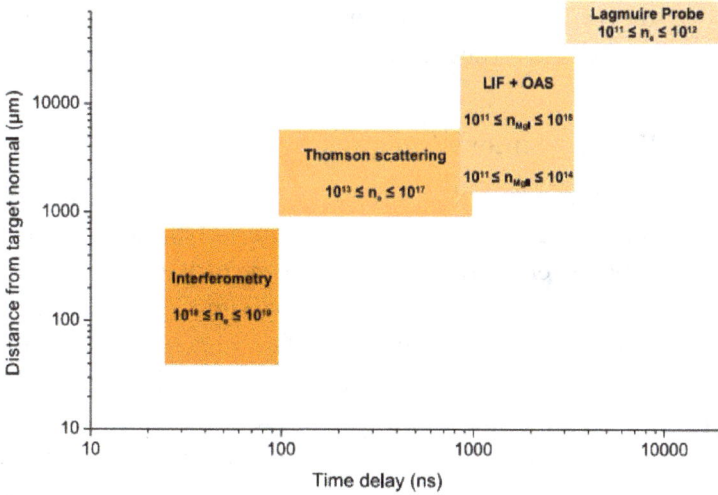

Figure 2.1. A schematic representation of the main investigation techniques with respect to their spatial and temporal applicability related to the typical range of detection limits (number densities in m^{-3}).

quality of the deposited film (Schou, 2009). On the other hand, the study of plasma phenomena, such as plume reflection, can help the development of PLD as a deposition technique by understanding some of its drawbacks in terms of changes in film stoichiometry and uniformity. Thus, we believe that a better understanding of the fundamental aspects of the LPP dynamics could allow us to control and improve the deposition process for the tailoring of new thin films with desirable properties.

In general, a given study presents results from only one technique (either optical or electrical). The idea of complementary methods used in tandem or simultaneously is rather rarely seen. In the first part of this chapter, we focus on the theoretical background of the main investigation techniques used during the preparation of this thesis, which fall into two categories: optical techniques (space- and time-resolved optical emission spectroscopy, Intensified Coupled Charge Device (ICCD) fast camera imaging) and electrical techniques (space- and time-resolved Langmuir probe measurements). In the second part of the chapter, we describe the details of the experimental setup from the University of Lille 1, which was used for the preparation of the thesis.

2.1. ICCD fast camera imaging

The ICCD fast camera photography technique (Janesick, 2013) is used to acquire bi-dimensional images of the global (i.e. not spectrally dispersed) optical emission of LPP. The recording of the images is usually performed using short integration times (the gate width of the ICCD camera) on the order of a few nanoseconds during the early stages of expansion up to tens or hundreds of nanoseconds at later stages. This particular technique allows the recording of LPP emission at various moments in time during its expansion and offers information about the global dynamics of the plasma, the spatial distribution and the structure of the ablated cloud. By automatically incrementing the image recording moment with respect to the ablation laser pulse, one can build a "movie" of the plasma emission evolution in time. We note however that each recorded image will contain only the projection of the plume emission on the CCD detector plane (e.g. xOy in Figure 2.2).

A CCD device usually contains a photoactive region and a shift register for the transfer of the collected data. When exposed to

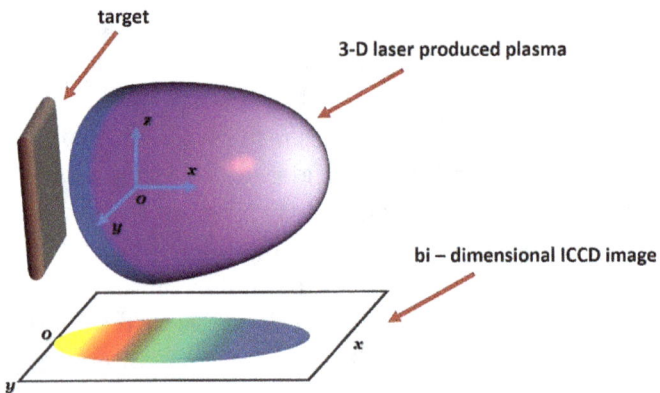

Figure 2.2. Conceptual representation of bi-dimensional images recording using ICCD fast camera photography technique.

photons, each pixel (n-type semiconductor) will generate an electric current proportional to the intensity of the light shed on it. The current collected by the CCD array will then be converted to a voltage, which is then digitized and stored. The intensifier attached to the CCD detector has the role of significantly amplifying the signal and thus improving the detection limit (up to 10^6-fold). The main components of the intensifier (placed in front of the CCD detector) are as follows: a photocathode, a microchannel and a phosphor screen. The microchannel plate (MCP) usually consists of a thin sheet of glass tubes ($\sim100\,\mu m$ in diameter) with length-to-diameter ratios of ~100. Gating is provided by applying a high voltage $\sim15\,kV$ between the front and back sides of the MCP. When the voltage is applied and a photon strikes on the photocathode, the extracted electron will further travel via multiple reflections down a thin channel in the MCP. Each tube has basically a continuous dynode structure resembling a photomultiplier. Since the electron is accelerated by the applied voltage, it gains enough kinetic energy and frees other electrons from the channel wall as it travels along it. The result of a single electron passing through the channels is $\sim10^3$ electrons. In the end, the electrons are striking a luminescent phosphorous screen which is placed at the back end of the system. The result is a significantly amplified optical signal which will be detected by the CCD and further transferred to the PC for analysis. The ICCD technique has seen rapid development in the last few years, and nowadays, there are available systems with a high resolution of 2048×2048 pixels, high frame rate up to $8\,MHz$ and a good signal-to-noise ratio. We note that despite the high frame rates available, a good temporal resolution of our "movie" mentioned above can be achieved only by recording each snapshot on successive ablation laser pulses. This requires good pulse-to-pulse stability of the laser ablation/plasma formation process.

Fast camera imaging is suitable for the investigation of transient phenomena. In the case of laser-produced plasmas, it is necessary to have an adequate triggering system and LPP generally has a lifetime of a few microseconds. Each recorded image is generally described by a series of parameters: resolution (which is given by the CCD detector and the optical system), time delay (the moment of time, with respect to the trigger signal, at which acquiring starts), and the "gate width" (or integration time). In our case, the initial moment ($t = 0$) is considered to be the "laser beam–target interaction moment". In order to have a good temporal resolution, the gate width is usually of a few nanoseconds and it can increase toward longer evolution time where the plume is more rarefied and the emission is weak. An example of a typical Mn plasma produced by nanosecond laser ablation in vacuum is given in Figure 2.3.

After the images are recorded, they are transferred to the computer where they can be further analyzed. The first step is a "pixel-to-cm" conversion that is based on simple calculations of the optical system used to image the plume on the ICCD detector. The second step is to perform a cross-section on the main expansion axis (Ox), thus allowing us to perform studies regarding the structure of the plasma. Each maximum in the emission cross-section describes a plasma component (Harilal *et al.*, 2003; Geohegan *et al.*, 1992).

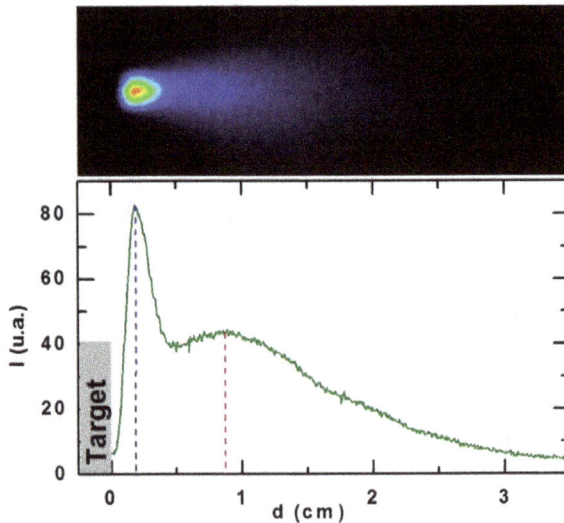

Figure 2.3. ICCD image of a nanosecond laser ablation Mn plasma (gate width: 10 ns, gate delay: 200 ns) and the corresponding cross-section on the plume expansion axis *Ox*.

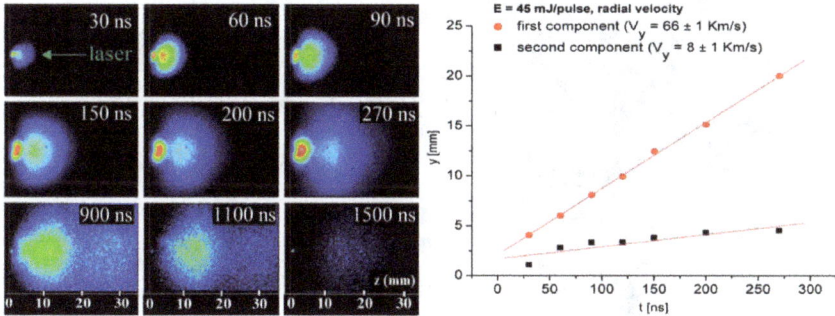

Figure 2.4. Temporal evolution of the global emission of a nanosecond laser-produced plasma on Al_2O_3 target (left) and the space–time displacement of the two emission maxima describing the two plasma components (right) (Ursu *et al.*, 2010).

By recording successive images of the plasma and representing the displacement of the emission maxima as a function of the moment of time at which the image was recorded (i.e. the delay), we can estimate the expansion velocity for each plasma component. An example from the literature (C. Ursu *et al.*, 2010) is given in Figure 2.4. Here, we can observe the increase in the volume of the plume as well as a decrease of the emitted light as the plume expands. From the evolution of the optical signals recorded by ICCD imaging, one can define and calculate various velocities (front, center-of-mass, radial) of the plasma structures (Ursu *et al.*, 2010).

Under vacuum and ultra-vacuum conditions, the laser-produced plasmas expand with a constant velocity, which can be extracted from the linear fit of the $d(t)$ representation (right-hand side of Figure 2.4), while in the presence of a background gas, the expansion can be described more accurately by a shockwave model (Geohegan and Puretzky, 1996) or by a drag model (Harilal *et al.*, 2003). In the literature, we can find other successful usages of this technique, especially considering the effect of the background pressure, laser wavelength or fluence on the plume expansion for both pure and complex targets (Borowitz *et al.*, 1987; Harilal *et al.*, 2003; Puretzky *et al.*, 1993). These studies led to the observation of other peculiar effects, besides the plume splitting in two structures,

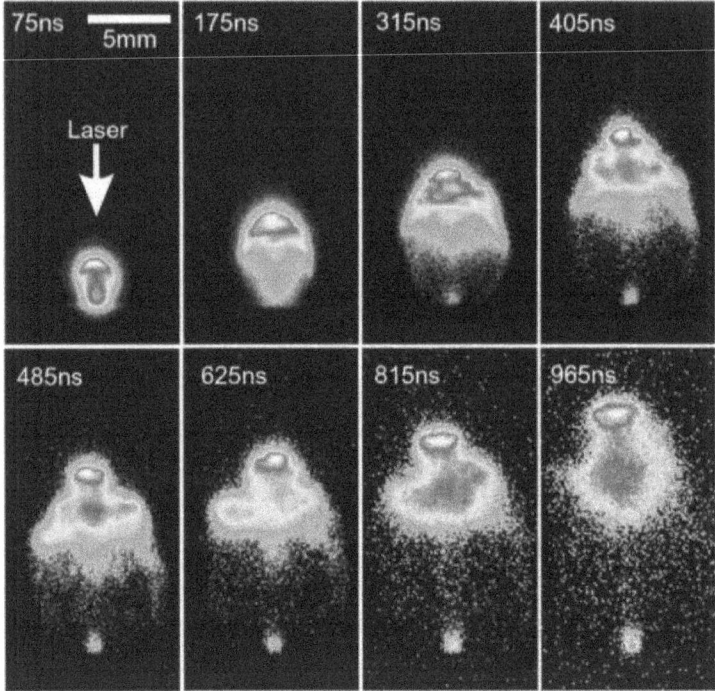

Figure 2.5. ICCD images of laser-produced plasma on an Aluminum target in 1.3 Torr background air pressure (Harilal *et al.*, 2003).

under vacuum conditions (Harilal, 2002; Gurlui and Focsa, 2011) and in the presence of a background gas (Harilal *et al.*, 2003; Puretzky *et al.*, 1993), the presence of a third plasma component containing mainly nanoparticles and clusters (Figure 2.5) (Geohegan and Puretzky, 1996; Harilal *et al.*, 2003). Of course, the limitation of this technique is its inability to differentiate between the contribution of various species and their spatial distribution within a complex plasma, as we collect all the emitted light from the plasma (i.e. not spectrally dispersed). This can be overcome by using specific band pass optical filters which allow the observation of only one type of species. The advantage of this approach was used by Bator *et al.* (2012), Canulescu *et al.* (2009) and Schou (2009), as they investigated the kinetic behavior of individual components and their spatial distribution during the deposition of thin films and correlated their stoichiometry differences with the individual properties of the elements and the overall properties of the complex targets.

The strength of a diagnosis process, with respect to this topic, consists in the ability to investigate as many aspects of the laser ablation plasma as possible in the same experimental conditions and setup configurations. As a complementary technique to the ICCD fast camera imagining in the literature, the most widespread is the optical emissions spectroscopy. This technique will delve into the dynamics of individual particles and the internal energy of the plasma plume. In the following section, we present the general aspects of the technique and its relevance to the laser ablation topic.

2.2. Optical Emission Spectroscopy

Optical emission spectroscopy (OES) consists in collecting the light emitted by the plasma and transferring it to the detector through a dispersive system. This technique is well established for steady plasma discharges (Bourg *et al.*, 2002). A more suitable version for the investigation of LPP and its transient nature is the space- and time-resolved OES, which translates as recording the emitted light at various moments in time with respect to the laser pulse and at various distances with respect to the target surface. The spatial resolution is here defined by the optical system preceding the dispersing system, while the temporal resolution is given by the ICCD gate. The collected emission spectrum of LPP can be (spectrally speaking) continuous (in the vicinity of the target and at short delays after the laser pulse (Singh and Thakur, 2006)) or discrete (at larger space–time coordinates (Harilal *et al.* 2013)). The discrete spectrum is composed of a series of emission lines that represent transitions between energetic discrete levels characteristic of every type of excited particles present in the plasma (see example in Figure 2.6). The spectral emission lines are characterized by three important parameters: wavelength, intensity and profile.

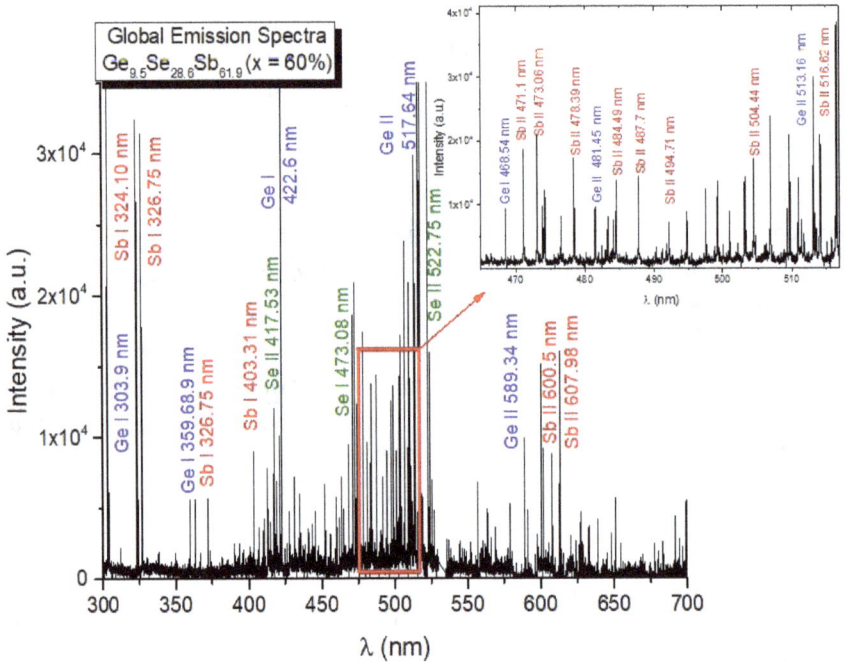

Figure 2.6. An example of the discrete emission of a $(GeSe_2)_{40}(Sb_2Se_3)_{60}$ laser ablation plasma at 6 mm from the target (25 ns gate delay, 2μs gate width) (Irimiciuc *et al.*, 2017a).

The optical emission spectroscopy technique can help determine the nature of the ejected particles through the energetic levels by identifying the wavelength and by using specialized databases (Kramida *et al.*, 2014. The profile and intensity of the spectral lines can also provide information regarding the interactions between the ejected particles (e.g. Stark broadening (Tankosić *et al.*, 2001; Cremers *et al.*, 2006) and the internal energy of the plasma (in the form of electron temperature and electron density), respectively.

Before discussing some quantitative aspects of the technique and describing the main plasma parameters that can be determined through OES, we must present some basic information about an important aspect connected to the measurement process: the thermodynamic equilibrium.

2.2.1. *Thermodynamic equilibrium in laser-produced plasmas*

In its most basic definition, a plasma is considered to be at thermodynamic equilibrium if all its temperatures are equal ($T_e = T_{\text{ex}} = T_i = T_{\text{rot/vib}}$). This is a strong restriction because if such a relation is satisfied, it will mean that there is an equal distribution of energy in the plasma volume (the excitation processes are equal to the thermal/kinetic ones and the rotational/vibrational movement) which is rarely the case. Thus, the "global" thermodynamic equilibrium can in principle never be reached in the case of "laboratory plasmas" due to the radiative disequilibrium (the de-excitation rate is higher than the excitation one). Therefore, the existence of a local thermodynamic equilibrium (LTE) is investigated for LPPs. The transient nature of LPP implies a space–time dependence of all plasma parameters, which means that the conditions for the LTE existence can be satisfied only for limited space–time ranges.

There are some general conditions used to estimate the local thermodynamic equilibrium. One of them relates to the electron density and states that LTE is reached if this plasma parameter is above a certain threshold. This is expressed by the McWhirter criterion (Fujimoto *et al.*, 1990; Cristoforetti *et al.*, 2013):

$$N_e(\text{cm}^{-1}) \geq 1.6 \times 10^{12} \Delta E^3 (\text{eV}) T_e^{\frac{1}{2}} (K), \qquad (2.1)$$

where N_e is the electron density, ΔE is the difference between the highest and the lowest atomic/ionic energy levels considered in the analysis, and T_e is the electron temperature.

Although the McWhirter criterion is a widespread tool in estimating the LTE, it is not a sufficient condition (Cristoforetti *et al.*, 2010) to completely establish its existence (i.e. for a lower range of electron densities, the corona equilibrium conditions can be applied (Cristoforetti *et al.*, 2013)). The criterion shows that LTE can be more easily achieved in cases of relatively dense plasmas, which is not always the case for LPP, where the density has a strong space–time dependence. Moreover, in the majority of plasma models used to describe the main plasma parameters, the plume is considered quasi-stationary. This means that within the investigated plasma volume, the spatial gradients are sufficiently small, so the effects induced by diffusion are neglected. For the case of partly ionized cold plasmas, as is the case of LPP, the diffusion process of atoms and ions represents again a strong restriction for the establishment of LTE. Therefore, as LPP has a strong space–time evolution, the LTE is often verified in the proximity of the target (a few millimeters from the target and at short delays after the laser pulse).

The LTE model is used to determine a series of plasma parameters like excitation temperature or electron density. In the framework of this model, the presence of LTE assumes an equilibrium between all energies inside the plasma (i.e. the electronic temperature is equal to the excitation one). This statement is not particularly true, as it will be shown in the following chapters due to the fact that not all the thermal energy of the plume is transferred to excitation processes. Assuming an equilibrium (Boltzmann) distribution function for the excited states, the intensity characterizing an excitation line can be described as

$$I_{ki} = N_0 \frac{hcA_{ki}g_k}{4\pi Z(T)\lambda} \exp\left(-\frac{E_k}{k_B T_e}\right), \qquad (2.2)$$

where N_0 is the total number density of atoms (particles), λ is the transition wavelength, A_{ki} is the Einstein coefficient of the k-i transition, g_k the statistical weight of the upper level, E_k is the energy of the upper level, $Z(T)$ is the partition function (Lochte-Holtgreven et al., 2000), h is the Planck constant, k_B is the Boltzmann constant and c is the light speed.

2.2.2. *Electron density and plasma excitation temperature*

Once it is established that the LTE model can be applied, the electron density can be estimated from the Saha-Eggert equation (Griem, 2005). The relationship connects the plasma ionization equilibrium temperature to the proportion of population of two successive ionization states. The "simplest" case is that of a neutral and a singly charged ion of the same species (Irissou *et al.*, 2002):

$$n_e = 4.83 \cdot 10^{15} \frac{I^* g^+ A^+ \lambda^*}{I^+ g^* A^* \lambda^+} T_e^{1.5} e^{-\frac{V^+ + E^+ - E^*}{k_B T_e}}, \tag{2.3}$$

where the $(^*, ^+)$ superscripts represent the neutral excited atom and the singly charged ion, respectively, I is the emission intensity of a spectral line of λ wavelength, T_e is the electron temperature (expressed in K), which is taken as the excitation temperature in LTE conditions, V^+ is the first ionization potential, and E is the energy of the upper level of the transition.

The electron temperature (which in the case of LTE is equal to the excitation temperature) can be simply calculated from (2.4) using the intensity ratio of two spectral lines emitted from the upper levels (E_1 and E_2) characterizing the same species (ion or atom):

$$T_e = \frac{E_2 - E_1}{k_b \ln \left(\frac{I_1 g_1 \lambda_1 f_1}{I_2 g_2 \lambda_2 f_2} \right)}, \tag{2.4}$$

where f_1 and f_2 are the oscillator strengths of the two spectral lines.

The spectroscopic data (E, A, f) can be found in various databases (Kramida *et al.*, 2014). We note however that there are some reserves concerning these values (especially oscillator strengths), which can lead to significant uncertainties. In order to minimize the errors, it is suitable to use not two but a series of atomic lines with different upper excitation levels. The Boltzmann plot method represents the logarithmic function of the line intensity versus the upper level energy:

$$\ln \left(\frac{I_{ki} \lambda}{g_k A_{ki}} \right) = \ln \left(N_0 \frac{hc}{4\pi Z(T)} \right) - \frac{E_k}{k_b T_e}. \tag{2.5}$$

The slope of this representation will give the excitation temperature, and its linearity or the deviation from it is considered as an indication of LTE validity (an example can be seen in Figure 2.7).

To conclude this brief presentation of the optical techniques used, we would like to mention that the space- and time-resolved OES can provide the spatial and temporal evolution of individual excited species (leading to the determination of the expansion velocities for

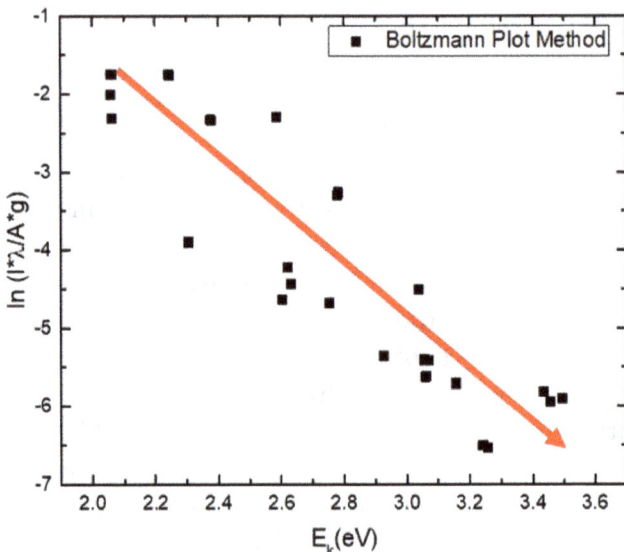

Figure 2.7. Example of Boltzmann plot obtained for a nanosecond laser-produced plasma on an Mg target; spectrum collected from a 0.3 mm width plasma slice situated at $d = 1$ mm from the target with $2\,\mu$s ICCD gate width.

atoms and ions) as well as their respective excitation temperature (Geohegan *et al.*, 1995; Ursu *et al.*, 2010; Harilal *et al.*, 1994). A major aspect that can be observed in the literature is that the excitation temperature is not uniform among different species (e.g. ions present different excitation temperatures compared to the corresponding atoms (Irimiciuc *et al.*, 2017a)). This aspect will be discussed in the following chapter, where we will present the evolution of the excitation temperature for a wide range of plasmas. Other important results reported in the literature showed that the emitted ions expand with a higher velocity (104 m/s) than the corresponding atoms (103 m/s), with about one order of magnitude difference. Moreover, Geohegan's group performed some systematic OES and ICCD studies to understand the complex plasma chemistry that takes place during the expansion of a carbon plasma in various background gases (Puretzky *et al.*, 1993, 2000a, 2000b), which revealed that complex molecules form during the expansion, with symmetrical distribution with respect to the plume axis.

2.3. Langmuir probe method for laser ablation plasmas

Historically, the Langmuir probe (LP) method was first proposed by Langmuir (Mott-Smith and Langmuir, 1926) in order to facilitate the description of ionized gases. Both the theory and the technical aspects surrounding this experimental technique have evolved over time (Lahm, 1974; Koopman, 1971; Merlino, 2007; Doggett *et al.*, 2009; Donnelly *et al.*, 2010a). Nowadays, the Langmuir probe presents great versatility, being used on various types of plasmas based on different technologies (laser-produced plasmas, discharge plasma, fusion plasmas or sputtering plasmas) and it presents itself in various configurations (Ovsyannikov *et al.*, 2000; Chen, 2003; Tang *et al.*, 2012) (single probe: plane, cylindrical, spherical (Figure 2.8); double probe or triple probe), being accepted as one of the major techniques for plasma investigations. The dynamics of the plasma particles (ions and electrons) in the vicinity of the probe does not differ fundamentally from one probe configuration to another,

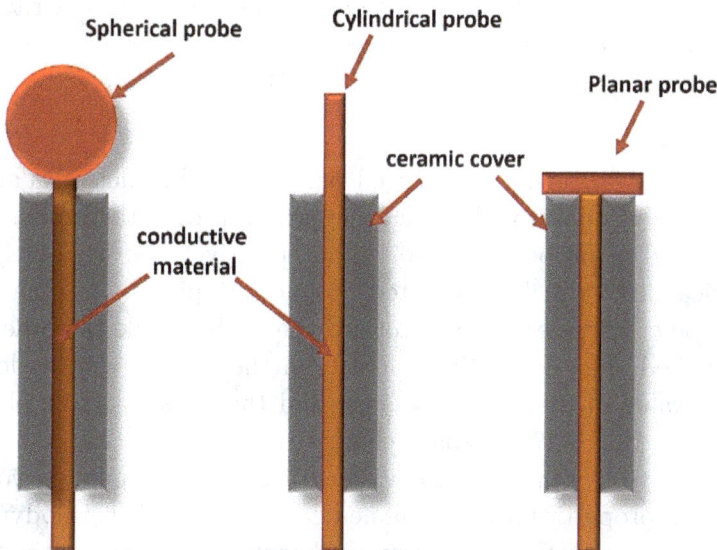

Figure 2.8. An illustration of the different types of Langmuir probes.

consequently no major difference will appear in the LP theory variants corresponding to the different geometrical configurations. In the following, we briefly describe the dynamics of the electrons and ions in the vicinity of the probe and how this can lead to the determination of a series of basic plasma parameters (temperature, density, plasma potential and velocity).

Let us consider, for now, the case of an ideal stationary plasma (i.e. neutral from an electric point of view, homogeneous and presenting local or global thermodynamic equilibrium). When the probe (a metallic electrode, for simplicity, we consider it plane) is immersed inside the plasma, the electrons, ions and atoms will arrive to the probe due to their thermal movement. Given the difference between the masses of the electrons and the ions, the amount of electrons per time unit that will arrive to the probe will be larger than that of ions per time unit and thus will negatively charge the probe surface. As a result, a surface is formed where the electrons will be rejected, while the ions will be accelerated toward the probe. Within this surface, the plasma neutrality is broken and there are no secondary ionizations. In this stationary regime, the overall number of electrons will be equal to that of the ions. This can be written as the equality between the two fluxes:

$$j_e = \frac{1}{4}en_ev_e = j_i = \frac{1}{4}en_iv_i, \qquad (2.6)$$

with n_e and n_i being the charged particle number densities and v_e and v_i the velocities with which the particles reach the probe.

When the probe is biased, either positively or negatively, an electrical field will be generated around the probe. As such, only one type of particle will pass toward the probe, while the other will be repelled. The "collecting" area from the plasma will be defined by the value of the applied voltage and the density of the plasma. Generally, this is called space charge surface, and it can spread on a few Debye lengths. The Debye length defines a minimum volume for which properties like plasma neutrality and local thermodynamic equilibrium are satisfied. The space charge is described by the Child relation $D = \lambda_{\text{Debye}} \frac{\sqrt{2}}{3} \left(\frac{2\,\text{eV}}{k_b T_e} \right)^{\frac{3}{4}}$, where D describes the thickness of the plasma sheet where the electron density is negligible, and $\lambda_{\text{Debye}} = \sqrt{\frac{k_b T_e \varepsilon_0}{n_e e^2}}$. The role of the applied voltage is not only to differentiate between different types of charges but also to separate them based on their energy. Thus, by sweeping a wide enough range of bias values, we will be able to collect all electrons and ions. If we take into account the simplest case, that of the planar configuration, the condition necessary to extract a particle from the plume is that the kinetic energy of the particle (the component oriented toward the probe surface) is higher than the space charge surface field $e(V_{\text{Plasma}} - V_{\text{Probe}})$, with V_{Plasma} the plasma potential and V_{Probe} the voltage applied on the probe. Therefore, the nature of the collected charge (and thus the probe current) will be dictated by the polarization of the target. For a positive potential, all the electrons will be collected, and due to their Maxwell–Boltzmann velocity distribution, the final relationship describing the electron current is

$$I_{\text{Probe}} = I(V_{\text{Probe}}) = I_e - I_i = I_{e0} \exp\left[\frac{-e(V_{\text{Plasma}} - V_{\text{Probe}})}{k_B T_e}\right] - I_{i0},$$

$$V_{\text{Probe}} < V_{\text{Plasma}}, \tag{2.7}$$

while for a negative potential,

$$I_{\text{Probe}} = I_{e0} - I_{i0} \exp\left[\frac{-e(V_{\text{Probe}} - V_{\text{Plasma}})}{k_B T_i}\right], \quad V_{\text{Probe}} > V_{\text{Plasma}}, \tag{2.8}$$

where I_{Probe} is the current collected by the probe, I_e and I_i are the electronic and the ionic currents, respectively, I_{e0} and I_{i0} are the respective saturation currents, e is the electronic charge, k_B is the Boltzmann constant, T_e and T_i are the electronic and ionic temperatures, respectively.

While the bias voltages are swept from high negative values to their corresponding positive ones, a characteristic similar to the ones presented in Figure 2.9 is recorded. This is called the I–V characteristic. Although the shape of the characteristic is slightly

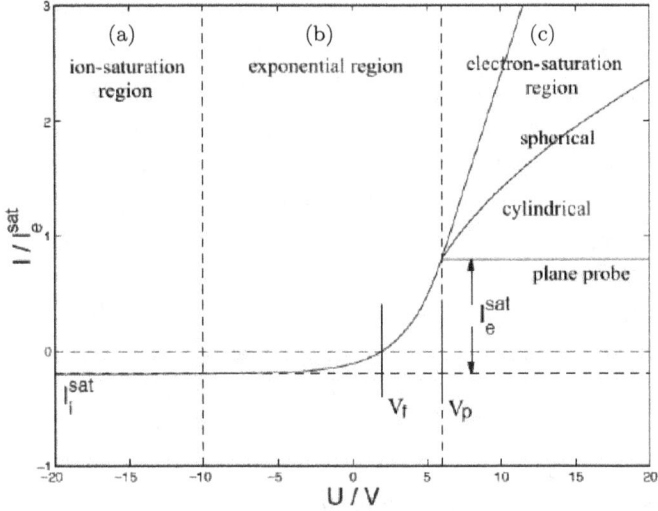

Figure 2.9. Typical I–V characteristics for various probe configuration (Bhattarai and Mishra, 2017; Chen, 1995).

dependent on the geometry of the probe, we can typically identify three different regions:

1. *The saturation ionic region defined by a small current amplitude and a relatively fast saturation for the ionic current*: The saturation current is defined as $I_{i0} = Aen_i\sqrt{\frac{T_e}{2m_i}}$, where A is the probe area, e is the electron charge, n_i is the ionic density, T_e is the electron temperature and m_i is the ionic mass.

2. *A transition part, where we identify an important point on the characteristic*: The floating potential (V_f), for which the current on the probe is null, is followed by an exponential increase in the electronic current. The inflection point of the characteristic, where the current changes from and exponential dependence on the V_{Probe} to a squared root one will define the plasma potential V_{Plasma}.

3. *The saturation electronic region defined by a maximum electronic current collected by the probe*: The saturation current is defined

as $I_{e0} = Aen_e\sqrt{\frac{T_e}{2\pi m_e}}$. This region is particularly characteristic of the planar probe, while for other configurations, the electron saturation is not reached. This is due to the increase of the space charge surface around the LP.

Once the $I-V$ characteristic is obtained, there are other parameters that can be identified. By reducing the ionic current and representing the evolution of the electron current as a function of the applied voltage in a logarithmic scale (Figure 2.10), we can further determine the electron temperature, the plasma potential and subsequently we can estimate the particle densities, the thermal velocities and the Debye length.

This technique can give good results for stationary plasmas, where by changing the position of the probe, one can map the properties of the plasma. A special case is attributed to transient plasmas, as is the case of LPP for which all properties present spatial and temporal evolution. This variation of the Langmuir probe theory for transient

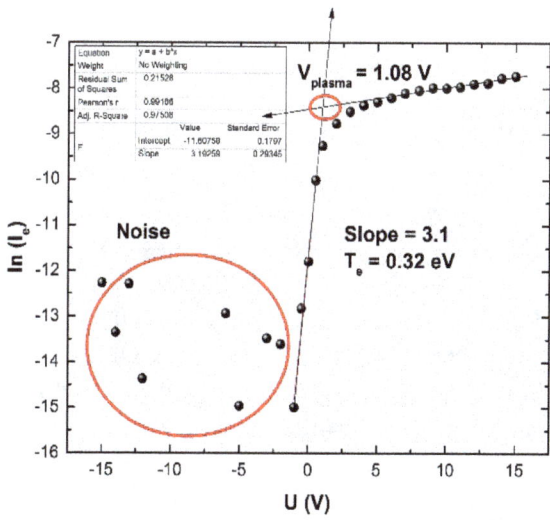

Figure 2.10. Example of semi-logarithmic representation of the electronic current (Irimiciuc *et al.*, 2017b).

plasmas was developed in the 1970s by Koopman (Koopman, 1971). For the case of laser-produced plasma, the current time-of-flight profiles $I = f(t)$ are representative of the velocity distribution of the ejected particles, through the relation $v = d/t$, where d is the target−probe distance and t is the arrival time of the particle at the probe. This is based on the assumption that the particle velocity is constant on the way from target to probe, which is supported (for LPP expansion under vacuum) by both experimental observations (Focsa *et al.*, 2009; S. Gurlui *et al.*, 2008; C. Ursu *et al.*, 2009) and theoretical considerations. According to Kelly's model (Roger Kelly and Dreyfus, 1988), in the proximity of the target, each particle will experience several collisions, leading to the formation of a Knudsen layer, followed by a supersonic expansion (Mihesan *et al.*, 2004). The Knudsen layer transforms the "half-range" ($v_z > 0$, with z-axis along the normal to the target) velocity distribution present at the sample surface into a "full-range" ($-\infty < v_z < +\infty$) Maxwell distribution superimposed on a drift velocity v_{drift}. In this hypothesis, the time dependence of the probe signal is given by (Zimmermann and Ho, 1995)

$$I_i(t) \propto \frac{1}{t^4} \exp\left[-\frac{m_i}{2k_b T_i} \left(\frac{d}{t} - v_{drift} \right)^2 \right]. \tag{2.9}$$

Equation (2.9) describes our experimental data well, as it can be seen in Figure 2.11, where the current TOF profile of a Cu plasma for a probe bias of $-10\,\text{V}$ was fitted.

This can be simplified even more, as it was shown in (Dogar *et al.*, 2011; Doggett and Lunney, 2009; Irimiciuc *et al.*, 2014), where it was considered that the movement of the charged particle "cloud" is defined by its "center-of-mass" velocity vCOM, derived as $d/t_{\max}$, where $t_{\max}$ (see Figure 2.11) is the moment at which the current reaches its maximum (this would correspond to the most probable velocity in a velocity distribution representation). The saturation ion current is then defined as

$$I_{i0} = eAn_i v_{\text{COM}} \tag{2.10}$$

These approaches, although they give important results, "ignore" the temporal evolution of the plasma parameters, in this case $T_{e,i}$, $n_{e,i}$,

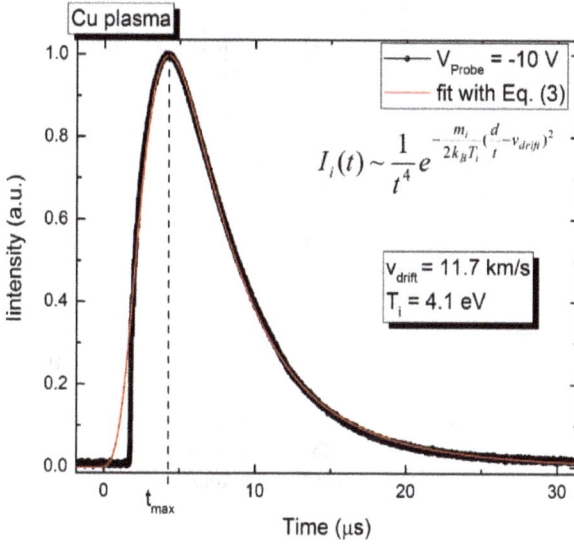

Figure 2.11. Example of ion temporal trace current of a femtosecond LPP on a Cu target ($V_{\text{Probe}} = -10\,\text{V}$) fitted with a shifted Maxwell–Boltzmann distribution (Irimiciuc *et al.*, 2017b).

considering instead an overall temperature and a constant drift velocity and a negligible thermal movement of the ejected particle. The situation can be salvaged by sampling the electronic and ionic temporal traces and various moments in time (Figure 2.12(a)). If the range of V_{Probe} covers both electronic and ionic saturation region, then we will be able to determine a series of plasma parameters ($T_{e,i}(t)$, $n_{e,i}(t)$, $V_{\text{Plasma}}(t)$, $V_{\text{Floating}}(t)$, $v_{\text{thermal}}(t)$ and $\lambda_{Debye}(t)$). Finding the saturation regions for LPP can be rather difficult as the properties of the plume depend strongly on the external parameters (laser fluence, background pressure, space–time coordinates). By assuming a v_{COM} of the order of $10^4\,\text{m/s}$ for the ejected particles (confirmed through ICCD fast camera photography (Ursu, 2009, 2010; Mihaila *et al.*, 2010; Pompilian, 2013)), this will mean that in order to reach saturation a minimum V_{Probe} of approximately $\pm 20\,\text{V}$ is necessary.

In Figure 2.12 are represented a series of characteristic signals for an Al plasma and the reconstructed I–V characteristics for two

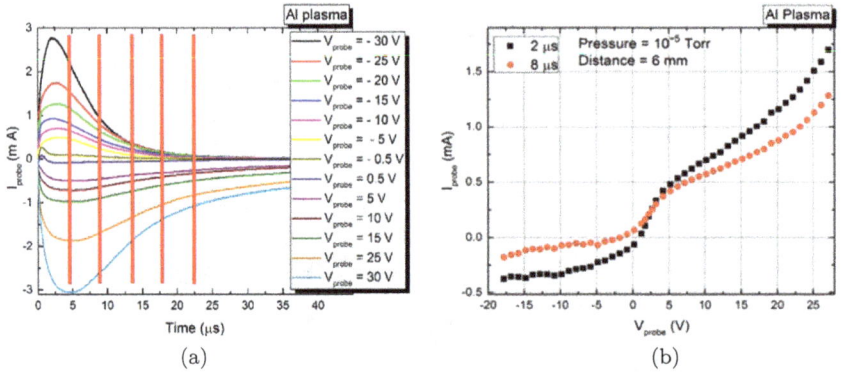

Figure 2.12. Aluminum laser-produced plasma: ionic and electronic currents collected at 6 mm from the target for various probe biases (a) and the reconstructed $I-V$ characteristics at two different delays (Irimiciuc *et al.*, 2017b) (b).

time delays. If the $I-V$ characteristic can be reconstructed at each moment of time, this allows us to treat the laser ablation plasma at each specific moment as being a homogeneous, stationary plasma, having a local thermodynamic equilibrium. Given these assumptions, some clear limitations arise. At short evolution times and small distances, the active surface is too large in comparison to the plasma volume and theoretical results are no longer valid since the probe shouldn't influence the plasma around it. Consequently, most LP measurements (Doggett and Lunney, 2009; Donnelly *et al.*, 2010a) are performed at long distances (typically a few centimeters) and at long evolution times (after 1 μs). This is understandable as the plume does increase its volume as it evolves, while the collecting area remains constant. Another important aspect which will be exploited in the following chapters is the influence that the probe has on the plasma surrounding it and how one can choose the optimal parameters for the electrical investigations.

Of course, there are other complimentary techniques that can be implemented. These have been discussed and used in a considerable number of papers, book chapters and presented at numerous conferences. Our goal here is not to go through every diagnostic tool available but only to showcase complimentary techniques that could easily be implemented in a synchronous manner on an industrial PLD setup as completed with feedback tools. In the following chapters, we show that the information extracted with these two techniques is potentially enough to completely describe the laser-produced plasmas during the deposition process.

References

Acquaviva S., Caricato A. P., De Giorgi M. L., Dinescu G., and Perrone A. 1997. Evidence for CN in spectroscopic studies of laser-induced plasma during pulsed irradiation of graphite targets in nitrogen and ammonia Evidence for CN in spectroscopic studies of laser-induced plasma during pulsed irradiation of graphite targets in nitrog, *J. Phys. B At. Mol. Opt. Phys.*, 30, 4405–4414.

Amoruso S., Schou J., Lunney J. G., and Phipps C. 2010. Ablation Plume Dynamics in a Background Gas, 665–676.

Anoop K. K., Polek M. P., Bruzzese R., Amoruso S., and Harilal S. S. 2015. Multidiagnostic analysis of ion dynamics in ultrafast laser ablation of metals over a large fluence range, *J. Appl. Phys.*, 117(8).

Bator M., Hu Y., Esposito M., Schneider C. W., Lippert T., and Wokaun A. 2012. Composition and species evolution in a laser-induced LuMnO3 plasma, *Appl. Surf. Sci.*, 258(23), 9355–9358.

Bhattarai S. and Mishra L. N. 2017. Theoretical study of spherical Langmuir probe in Maxwellian plasma, *Int. J. Physics*, 5(3), 73–81.

Boné A., Lemos N., Figueira G., and Dias J. M. 2016. Quantitative shadowgraphy for laser–plasma interactions, *J. Phys. D. Appl. Phys.*, 49(15), 155204.

Borowitz J. L., Eliezer S., Gazit Y., Givon M., Jackel S., Ludmirsky A., D. Salzmann, Yarkoni E., Zigler A., and Arad B. 1987. Temporally resolved target potential measurements in laser-target interactions, *J. Phys. D. Appl. Phys.*, 20(2), 210–214.

Canulescu S., Papadopoulou E. L., Anglos D., Lippert T., Schneider C. W., and Wokaun A. 2009. Mechanisms of the laser plume expansion during the ablation of LiMn[sub 2]O[sub 4], *J. Appl. Phys.*, 105(6), 63107.

Chen F. F. 1995. *Introduction to Plasma Physics*, Springer, US.

Chen F. F. 2003. *Lecture Notes on Langmuir Probe Diagnostics*.

Chen J., Lippert T., Ojeda-G-P A., Stender D., Schneider C. W., and Wokaun A. 2014. Langmuir probe measurements and mass spectrometry of plasma plumes generated by laser ablation of La0.4Ca0.6MnO3, *J. Appl. Phys.*, 116(7), 73303.

Cremers D. A. and Radziemski L. J. 2006. *Handbook of Laser-Induced Breakdown Spectroscopy*, John Wiley & Sons Ltd, Oxford, UK.

Cristoforetti G., Legnaioli S., Pardini L., Palleschi V., Salvetti A., and Tognoni E. 2006. Spectroscopic and shadowgraphic analysis of laser induced plasmas in the orthogonal double pulse pre-ablation configuration, *Spectrochim. Acta Part B At. Spectrosc.*, 61(3), 340–350.

Cristoforetti G., Tognoni E., and Gizzi L. A. 2013. Thermodynamic equilibrium states in laser-induced plasmas: From the general case to laser-induced breakdown spectroscopy plasmas, Spectrochim. *Acta - Part B At. Spectrosc.*, 90, 1–22.

Dogar A. H., Ilyas B., Ullah S., Nadeem A., and Qayyum A. 2011. Langmuir probe measurements of Nd-YAG laser-produced copper plasmas, *IEEE Trans. Plasma Sci.*, 39, 897–900.

Doggett B. and Lunney J. G. 2009. Langmuir probe characterization of laser ablation plasmas, *J. Appl. Phys.*, 105(3), 33306.

Donnelly T., Lunney J. G., Amoruso S., Bruzzese R., Wang X., and Ni X. 2010a. Dynamics of the plumes produced by ultrafast laser ablation of metals, *J. Appl. Phys.*, 108(4).

Donnelly T., Lunney J. G., Amoruso S., Bruzzese R., Wang X., and Phipps C. 2010b. Plume dynamics in femtosecond laser ablation of metals, 643(2010), 643–655.

Doria D., Lorusso A., Belloni F., and Nassisi V. 2004. Characterization of a nonequilibrium XeCl laser-plasma by a movable Faraday cup, *Rev. Sci. Instrum.*, 75(2), 387.

Focsa C., Nemec P., Ziskind M., Ursu C., Gurlui S., and Nazabal V. 2009. Laser ablation of AsxSe100–x chalcogenide glasses: Plume investigations, *Appl. Surf. Sci.*, 255(10), 5307–5311.

Focsa C., Gurlui S., Nica P., Agop M., and Ziskind M. 2017. Plume splitting and oscillatory behavior in transient plasmas generated by high-fluence laser ablation in vacuum, *Appl. Surf. Sci.*, 424, 299–309.

Fujimoto T. 1990. Validity criteria for local thermodynamic equilibrium in plasma spectroscopy, *Phys. Rev. A - At. Mol. Opt. Phys.*, 42(11), 6588–6601.

Geohegan D. B. and Puretzky A. A. 1996. Laser ablation plume thermalization dynamics in background gases: Combined imaging, optical absorption and emission spectroscopy, and ion probe measurements, *Appl. Surf. Sci.*, 96–98, 131–138.

Griem H. R. 2005. *Principles of Plasma Spectroscopy*, Cambridge Univ. Press.

Gurlui S. and Focsa C. 2011. Laser ablation transient plasma structures expansion in vacuum, *IEEE Trans. Plasma Sci.*, 39(11), 2820–2821.

Gurlui S., Agop M., Nica P., Ziskind M., and Focsa C. 2008. Experimental and theoretical investigations of a laser-produced aluminum plasma, *Phys. Rev. E*, 78(2), 26405.

Hansen T. N., Schou J., and Lunney J. G. 1999. Langmuir probe study of plasma expansion in pulsed laser ablation, *Appl. Phys. A*, 69, S601–S604.

Harilal S. S., Issac R. C., Bindhu C. V., Nampoori V. P. N., and Vallabhan C. P. G. 1999. Optical emission studies of species in laser-produced plasma from carbon, *J. Phys. D. Appl. Phys.*, 30(12), 1703–1709.

Harilal S. S., Bindhu C. V., Tillack M. S., Najmabadi F., and Gaeris A. C. 2003. Internal structure and expansion dynamics of laser ablation plumes into ambient gases, *J. Appl. Phys.*, 93(5), 2380.

Harilal S. S., Miloshevsky G. V., Sizyuk T., and Hassanein A. 2013. Effects of excitation laser wavelength on Ly and He line emission from nitrogen plasmas, *Phys. Plasmas*, 20(1).

Harilal, S. S., Farid N., Freeman J. R., Diwakar P. K., LaHaye N. L., and Hassanein A. 2014. Background gas collisional effects on expanding fs and ns laser ablation plumes, *Appl. Phys. A*, 117(1).

Hussein A. E., Diwakar P. K., Harilal S. S. and Hassanein A. 2013. The role of laser wavelength on plasma generation and expansion of ablation plumes in air, *J. Appl. Phys.*, 113(14).

Irimiciuc S., Boidin R., Bulai G., Gurlui S., Nemec P., Nazabal V., and Focsa C. 2017a. Laser ablation of (GeSe2)100-x(Sb2Se3)x chalcogenide glasses: Influence of the target composition on the plasma plume dynamics, *Appl. Surf. Sci.*, 418, 594–600.

Irimiciuc S. A., Mihaila I., and Agop M. 2014. Experimental and theoretical aspects of a laser produced plasma, *Phys. Plasmas*, 21(9).

Irimiciuc S. A., Gurlui S., Bulai G., Nica P., Agop M., and Focsa C. 2017b. Langmuir probe investigation of transient plasmas generated by femtosecond laser ablation of several metals: Influence of the target physical properties on the plume dynamics, *Appl. Surf. Sci.*, 417, 108–118.

Irissou E., Le Drogoff B., Chaker M., and Guay D. 2002. Correlation between plasma expansion dynamics and gold-thin film structure during pulsed-laser deposition, *Appl. Phys. Lett.*, 80(10), 1716.

Janesick J. R. (ed.) 2013. *Scientific Charge-Coupled Devices*, SPIE, Washington.

Kelly R. and Dreyfus R. W. 1988. On the effect of Knudsen-layer formation on studies of vaporization, sputtering, and desorption, *Surf. Sci.*, 198(1–2), 263–276.

Kokai F., Takahashi K., Shimizu K., Yudasaka M., and S. Iijima 1999. Shadowgraphic and emission imaging spectroscopic studies of the laser ablation of graphite in an Ar gas atmosphere, *Appl. Phys. A Mater. Sci. Process.*, 69(7), 223–227.

Koopman D. W. 1971. Langmuir probe and microwave measurements of the properties of streaming plasmas generated by focused laser pulses, *Phys. Fluids*, 14(8), 1707.

Kramida A., Ralchenko Y., Reader J., and NIST ASD Team. 2014. NIST Atomic Spectra Database Lines Form, NIST At. Spectra Database (ver. 5.2).

Lahm S. H. 1901. Unified theory for the Langmuir probe in a collisionless plasma, *Phys. Fluids*, 8, 73, 1965.

Liu X., Du D., and Mourou G. 1997. Laser ablation and micromachining with ultrashort laser pulses, IEEE J. *Quantum Electron.*, 33(10), 1706–1716.

Lochte-Holtgreven W., North-Holland E., Ovsyannikov A. A., and Zhukov M. F. 2000. *Plasma Diagnostics*, Cambridge International Science Publishing.

Merlino R. L. 2007. Understanding Langmuir probe current-voltage characteristics, *Am. J. Phys.*, 75(12), 1078.

Mihaila I., Ursu C., Gegiuc A., and Popa G. 2010. Diagnostics of plasma plume produced by laser ablation using ICCD imaging and transient electrical probe technique, *J. Phys. Conf. Ser.*, 207, 12005.

Mihesan C., Lebrun N., Ziskind M., Chazallon B., Focsa C., and Destombes J. L. 2004. IR laser resonant desorption of formaldehyde-H_2O ices: Hydrated cluster formation and velocity distribution, *Surf. Sci.*, (1–3), 566–568.

Mott-Smith H. M. and Langmuir I. 1926. The theory of collectors in gaseous discharges, *Phys. Rev.*, 28(4), 727–763.

Němec P., Olivier M., Baudet E., Kalendová A., Benda P., and Nazabal V. 2014. Optical properties of (GeSe2)100-x(Sb 2Se3)x glasses in near- and middle-infrared spectral regions, *Mater. Res. Bull.*, 51, 176–179.

Olivier, M., Tchahame J. C., Němec P., Chauvet M., Besse V., Cassagne C., Boudebs G., Renversez G., Boidin R., Baudet E., and Nazabal V. 2014. Structure, nonlinear properties, and photosensitivity of (GeSe2)100-x(Sb2Se3)x glasses, *Opt. Mater. Express*, 4(3), 525–540.

Ovsyannikov A. A. and Zhukov M. F. 2000. *Plasma Diagnostics*, Cambridge International Science Publishing.

Pompilian O. G. 2013. *Pulsed Laser Deposition and Characterization of Chalcogenide Thin Films*, Université Lille 1 – Sciences Et Technologies.

Pompilian O. G., Gurlui S., Nemec P., Nazabal V., Ziskind M., and Focsa C. 2013. Plasma diagnostics in pulsed laser deposition of GaLaS chalcogenides, *Appl. Surf. Sci.*, 278, 352–356.

Puretzky A. A., Geohegan D. B., Haufler R. E., Hettich R. L., Zheng X. Y., and Compton R. N. 1993. Laser ablation of graphite in different buffer gases, *AIP Conf. Proc.*, 288(1993), 365–374.

Puretzky A. A., Geohegan D. B., Fan X., and Pennycook S. J. 2000a. Dynamics of single-wall carbon nanotube synthesis by laser vaporization, *Appl. Phys. A Mater. Sci. Process.*, 70(2), 153–160.

Puretzky A. A., Geohegan D. B., Fan, X., and Pennycook S. J. 2000b. In situ imaging and spectroscopy of single-wall carbon nanotube synthesis by laser vaporization, *Appl. Phys. Lett.*, 76(2).

Schou J. 2009. Physical aspects of the pulsed laser deposition technique: The stoichiometric transfer of material from target to film, *Appl. Surf. Sci.*, 255(10), 5191–5198.

Singh J. and Thakur S. 2006. *Laser Induced Breakdown Spectroscopy*, Elsevier.

Smithells C. J. 2004. Smithells metals reference book, W. F. Gale and T. C. Totemeier (eds.), Elsevier Butterworth-Heinemann.

Strickland, D. and Mourou, G. 1985. Compression of amplified chirped optical pulses, *Opt. Commun.*, 56(3), 219–221.

Sunil S., Kumar A., Singh R. K., and Subramanian K. P. 2008. Measurements of electron temperature and density of multi-component plasma plume formed by laser-blow-off of LiF-C film, *J. Phys. D. Appl. Phys.*, 41(8), 85211.

Tang E., Xiang, S., Yang, M., and Li, L. (2012), Sweep Langmuir probe and triple probe diagnostics for transient plasma produced by hypervelocity impact, *Plasma Sci. Technol.*, 14(8), 747–753.

Tankosić D., Popović L. Č., and Dimitrijević M. S. 2001. Electron-impact stark broadening parameters for ti ii and ti iii spectral lines, *At. Data Nucl. Data Tables*, 77(2), 277–310.

Ursu C. 2010. Caracterisation par methodes optiques et electriques du plasma produit par ablation laser, Université Lille 1 – Sciences Et Technologies.

Ursu C. and Nica P. 2013. Diagnosis of carbon laser produced plasma by using an electrostatic energy analyzer, *Opt. Adv. Mat.*, 15(1), 42–45.

Ursu C., Gurlui S., Focsa C., and Popa G. 2009. Space- and time-resolved optical diagnosis for the study of laser ablation plasma dynamics, *Nucl. Instrum. Methods Phys. Res. Sect. B Beam Interact. with Mater. Atoms*, 267(2), 446–450.

Ursu C., Pompilian O. G., GurluiS., Nica P., Agop M., Dudeck M. and Focsa C. 2010. Al_2O_3 ceramics under high-fluence irradiation: plasma plume dynamics through space- and time-resolved optical emission spectroscopy, *Appl. Phys. A*, 101(1), 153–159.

Weaver, I., Martin G. W., Graham W. G., Morrow T., and Lewis C. L. S. 1999. The Langmuir probe as a diagnostic of the electron component within low temperature laser ablated plasma plumes, *Rev. Sci. Instrum.*, 70(3).

Yao Y. L., Chen H., and Zhang W. 2005. Time scale effects in laser material removal: A review, *Int. J. Adv. Manuf. Technol.*, 26(5), 598–608.

Yeates P., Fallon C., Kennedy E. T., and Costello J. T. 2013. Atomic mass dependent electrostatic diagnostics of colliding laser plasma plumes, *Phys. Plasmas*, 20(9), doi:10.1063/1.4821979.

Zimmermann F. M. and Ho W. 1995. State resolved studies of photochemical dynamics at surfaces, *Surf. Sci. Rep.*, 22(4–6), 127–247.

Chapter 3

Toward a Theory of Motion in a Multifractal Paradigm

3.1. Generalities

In its standard form, the scale relativity theory (SRT) (Nottale, 1993, 2011) employs monofractal dynamics in order to describe complex system (biostructures, economic systems, etc. (Mandelbrot, 2006; Politi and Badii, 2003; Mitchell, 2011)) behaviors. These monofractal dynamics operate through monofractal curves, which are continuous and non-differentiable (characterized by a single fractal dimension D_F). Moreover, in the SRT model, a "privileged" fractal dimension exists, namely $D_F \rightarrow 2$: a situation in which the dynamics of any complex system are described on monofractal manifolds (either through Schrödinger-type geodesics or through hydrodynamic-type ones). In particular, for the dynamics (of complex system entities) of Peano-type curves (i.e. $D_F \rightarrow 2$) at Compton scale resolutions (Nottale, 1993, 2011), Schrödinger-type geodesics of a monofractal manifold can be identified with the standard Schrödinger equation of quantum mechanics (Nottale, 1993, 2011). In the same context, hydrodynamic-type geodesics of monofractal manifolds can be identified with the hydrodynamic model of quantum mechanics (Nottale, 1993, 2011). In such a conjecture, information, in its various forms (Marinescu, 2011), implies any type of monofractal manifold (Agop et al., 2014, 2015, 2016; Grigorovici et al., 2017):

(i) monofractal tensors of viscous stress, induced by non-differential components of velocity fields;

43

(ii) uncertainty relationships for constant values of Onicescu's informational energy (Alipou and Mohajcri, 2012; Agop *et al.*, 2015a) (values induced through the informational energy maximization principle and assimilated to transitivity manifolds of $SL(2R)$-type groups);

(iii) harmonic mappings from the usual space into the Lobachevsky plane through the scalar potentials of the velocity fields;

(iv) "correlation" of the kinetic moments (orbital, spin or orbital–spin) with "pairs" generation through invariant functions of $SL(2R)$-type groups.

Nature, however, is multifractal at any scale resolution, both from a structural and functional point of view (Mandelbrot, 2006; Politi and Badii, 2003; Mitchell, 2011; Weinberg, 1994; Hawking and Penrose, 1996; Argyris *et al.*, 2002; Penrose, 2007). Thus, it is necessary to extend the SRT in order to be able to analyze and operate it using multifractal dynamics. This is due to the fact that motion curves belonging to the entities of any complex structure are multifractal curves, i.e. they are continuous and non-differentiable, simultaneously characterized by several fractal dimensions and explicated by the singularity spectrum $f(\alpha)$ (with α being the singularity index (Mandelbrot, 2006)). To this end, we present in Section 3.2 the multifractalization procedure, as part of a "general methodology" of the SRT in the form of the multifractal theory of motion (Agop and Păun, 2017; Agop and Mercheş, 2019; Agop *et al.*, 2019; Gavriluţ *et al.*, 2019; Mazilu *et al.*, 2020). In such a conjecture, in Section 3.3, we analyze the dynamics of complex systems in the form of Schrödinger- and hydrodynamic-type "regimens" at various scale resolutions. Then, in the same framework, by accepting the principle of Shannon's information minimization, we show that, for complex system dynamics with radial symmetry, Newtonian-type behaviors dependent on scale resolution can be found (Section 3.4). The analysis of these behaviors by means of explaining motion geodesics in their conic-type analytical form, using a set of complex variables, specifies the fact that the center of the Newtonian-type multifractal force is different from the center of the multifractal trajectory, with the measure of this difference being the eccentricity, which depends on the initial conditions (Section 3.5). In the end (Section 3.6), the eccentricities' geometry, i.e. the initial conditions geometry, becomes, through the Cayley–Klein metrization principle, the Lobachevsky plane geometry. Then, the harmonic mappings between the usual space and the Lobachevsky plane in a Poincaré metric show that the Ernst potential of general relativity becomes, in essence, of a classical nature. Thus, the Newtonian-type multifractal dynamics, perceived and described through the multifractal theory of motion, become a local manifestation of the gravitational field of general relativity, as is required.

3.2. A word on the multifractal theory of motion

In what follows, we admit that the motions of the entities of any complex system are described by continuous and non-differentiable curves (multifractal curves). Such a "non-differentiable" procedure to approach these motions has important consequences (Nottale, 1993, 2011; Agop and Păun, 2017; Agop and Merches, 2019; Agop *et al.*, 2019; Gavriluţ *et al.*, 2019; Mazilu *et al.*, 2020):

(i) Any multifractal curve is explicitly scale δt dependent, i.e. its length tends to infinity when δt tends to zero (Lebesgue theorem) (Mandelbrot, 2006). Moreover, the space of any complex system dynamics becomes a multifractal in the Mandelbrot sense.

(ii) The dynamics of any complex system are related to the behavior of a set of functions during the zoom operation of δt, i.e. $\delta t \equiv dt$ through the functionality of the substitution principle.

(iii) The dynamics of any complex system are described through multifractal variables. Then, two derivatives of any variable field, $Q(t, dt)$, which describe the complex system dynamics, can be defined:

$$
\begin{aligned}
\left(\frac{dQ}{dt}\right)_+ &= \lim_{\Delta t \to 0} \frac{Q(t, t + \Delta t) - Q(t, \Delta t)}{\Delta t}, \\
\left(\frac{dQ}{dt}\right)_- &= \lim_{\Delta t \to 0} \frac{Q(t, \Delta t) - Q(t - \Delta t, \Delta t)}{\Delta t}.
\end{aligned}
\tag{3.1}
$$

The sign "+" corresponds to the forward dynamics, while the sign "−" corresponds to the backward ones.

(iv) The differential of the spatial coordinate field has the form

$$d_\pm X^i(t, dt) = d_\pm x^i(t) + d_\pm \xi(t, dt). \tag{3.2}$$

The differentiable part of X^i, i.e. $d_\pm x^i(t)$ is scale resolution independent. The non-differentiable part of X^i, i.e. $d_\pm \xi(t, dt)$ is scale resolution dependent.

(v) The non-differentiable part of the spatial coordinate field, which describes the complex system dynamics, satisfies the non-differentiable multifractal equation (Mandelbrot, 2006; Lakshmanan and Rajaseekar, 2003; Cristescu, 2008)

$$d_\pm \xi^i(t, dt) = \lambda^i_\pm (dt)^{\left[\frac{2}{f(\alpha)}\right] - 1}, \tag{3.3}$$

where $\lambda^i_\pm$ are constant coefficients associated with the differentiable–non-differentiable scale transition, $f(\alpha)$ is the singularity spectrum of order α of the fractal dimension D_F and α is the singularity index. One can find various definitions for fractal dimensions: the fractal dimension in the sense of Kolmogorov, the fractal dimension in the sense of Hausdorff–Besikovich, etc. (Mandelbrot, 2006). By selecting one of these definitions and operating in complex system dynamics, it is mandatory that the value of the fractal dimension be constant and arbitrary for the entirety of the dynamical analysis. Usually, $D_F < 2$ can be chosen for correlative processes, while $D_F > 2$ is used for non-correlative processes, etc. (Nottale, 2011). In such a conjecture, through (3.3), it is possible to identify not only the "areas" of the complex system dynamics that are characterized by a certain fractal dimension (i.e. in the case of monofractal dimensions) but also the number of "areas" in which fractal dimensions are situated in a values interval (i.e. in the case of multifractal dimensions). Furthermore, through the singularity spectrum $f(\alpha)$, we can identify classes of universality in the complex system dynamic laws, even when regular or strange attractors have different aspects (Mandelbrot, 2006; Lakshmanan and Rajaseekar, 2003).

(vi) The differential time reflection invariance of any variable is recovered by means of the operator

$$\frac{\hat{d}}{dt} = \frac{1}{2}\left(\frac{d_+ + d_-}{dt}\right) - \frac{i}{2}\left(\frac{d_+ - d_-}{dt}\right). \tag{3.4}$$

This is a natural result of Cresson's theorem (Cresson, 2003). For example, applying the operator (3.4) to X^i yields the complex velocity fields

$$\hat{V}^i = \frac{\hat{d}X^i}{dt} = V_D^i - V_F^i \tag{3.5}$$

with

$$V_D^i = \frac{1}{2}d_+X^i + \frac{d_+X^i}{dt}, \quad V_F^i = \frac{1}{2}d_+X^i - \frac{d_-X^i}{dt}, \quad i = 1,2,3. \tag{3.6}$$

The real part of $\hat{V}^i$, i.e. V_D^i (differential velocity) is scale resolution independent. The imaginary one V_F^i (non-differentiable velocity) is scale resolution dependent.

(i) Since the multifractalization describing complex system dynamics implies stochasticization (Mandelbrot, 2006; Lakshmanan and Rajaseekar, 2003; Cristescu, 2008), the whole statistic "arsenal" in the form of averages, variances, covariances, etc., becomes operational. Thus, let us choose for the average of $d_\pm X^i$, the following functionality:

$$\langle d_\pm X^i \rangle \equiv d_\pm x^i, \tag{3.7}$$

with

$$\langle d_\pm \xi^i \rangle = 0. \tag{3.8}$$

The previous relation (3.8) implies that the average of the non-differential part of the spatial coordinate field is null.

(vii) The complex system dynamics can be described through the scale covariant derivative given by the operator (Agop and Paun, 2017; Agop and Merches, 2019; Agop *et al*, 2019)

$$\frac{\hat{d}}{dt} = \partial_t + \hat{V}^i \partial_i + \frac{1}{4}(dt)^{\left[\frac{2}{f(\alpha)}\right]-1} D^{lk} \partial_l \partial_k, \qquad (3.9)$$

where

$$D^{lk} \partial_l = \left(\lambda_+^l \lambda_+^k - \lambda_-^l \lambda_-^k \right) + i \left(\lambda_+^l \lambda_+^k + \lambda_-^l \lambda_-^k \right),$$

$$\partial_t = \frac{\partial}{\partial t}, \partial_i = \frac{\partial}{\partial X^i}, \partial_l \partial_k = \frac{\partial^2}{\partial X^l \partial X^k}. \qquad (3.10)$$

For Markov-type stochastic processes (Agop and Paun, 2017), i.e.

$$\lambda_+^i \lambda_+^l = \lambda_-^i \lambda_-^l = 2\lambda \delta^{il} \qquad (3.11)$$

and

$$f(\alpha) \equiv D_F, \qquad (3.12)$$

where λ is a specific coefficient associated with the multifractal–non-multifractal scale transition and δ^{il} is Kronecker's pseudo-tensor, the scale covariant derivative becomes

$$\frac{\hat{d}}{dt} = \partial_t + \hat{V}^l \partial_l - i\lambda(dt)^{\left[\frac{2}{D_F}\right]-1} \partial_l \partial^l. \qquad (3.13)$$

In the particular case of Peano-type curves, which implies $D_F = 2$, the scale covariant derivative (3.13) takes the standard form from the SRT:

$$\frac{\hat{d}}{dt} = \partial_t + \hat{V}^l \partial_l - iD \partial_l \partial^l \qquad (3.14)$$

where $\lambda \equiv D$ is the diffusion coefficient associated with the fractal–non-fractal scale transition. Therefore, this model generalizes all the results of Nottale's theory (i.e. the SRT) (Nottale, 1993, 2011). Moreover, for Compton scale resolution, (3.14) becomes the "quantum operator" (see Nottale (1993, 2011)).

3.3. Dynamics in complex systems through Schrödinger- and hydrodynamic-type "regimens" at various scale resolutions

Now, accepting the functionality of the scale covariance principle (see Nottale (1993, 2011)), i.e. applying the operator (3.9) to the complex velocity fields (3.5), in the absence of any external constraint, the motion equation (the geodesics equation) takes the following form:

$$\frac{d\hat{V}^i}{dt} = \partial_t \hat{V}^i + \hat{V}^l \partial_l \hat{V}^i + \frac{1}{4}(dt)^{\left[\frac{2}{f(\alpha)}\right]-1} D^{lk} \partial_l \partial_k \hat{V}^i = 0. \tag{3.15}$$

This means that for any complex system dynamics, the multifractal acceleration, $\partial_t \hat{V}^i$, the multifractal convection $\hat{V}^l \partial_l \hat{V}^i$ and the multifractal dissipation $D^{lk}\partial_l\partial_k \hat{V}^i$ make their balance at any point of the multifractal curve. Particularly, for (3.11) and (3.12), the motion equation (the geodesics equation) (3.15) becomes

$$\frac{d\hat{V}^i}{dt} = \partial_t \hat{V}^i + \hat{V}^l \partial_l \hat{V}^i - i\lambda(dt)^{\left[\frac{2}{D_F}\right]-1} \partial_l \partial^l \hat{V}^i = 0. \tag{3.16}$$

Now, separating the complex system dynamics on scale resolutions (differentiable and non-differentiable) (3.15) becomes

$$\partial_t V_D^i + V_D^l \partial_l V_D^i - V_F^l \partial_l V_F^i + \frac{1}{4}(dt)^{\left[\frac{2}{f(\alpha)}\right]-1} D^{lk}\partial_l\partial_k V_D^i = 0,$$
$$\partial_t V_F^i + V_F^l \partial_l V_D^i + V_D^l \partial_l V_F^i + \frac{1}{4}(dt)^{\left[\frac{2}{f(\alpha)}\right]-1} D^{lk}\partial_l\partial_k V_F^i = 0, \tag{3.17}$$

while (3.16) takes the form

$$\partial_t V_D^i + V_D^l \partial_l V_D^i - \left[V_F^l + \lambda(dt)^{\left[\frac{2}{f(\alpha)}\right]-1} \partial^l \right] \partial_l V_F^i = 0,$$
$$\partial_t V_F^i + V_D^l \partial_l V_F^i + \left[V_F^l + \lambda(dt)^{\left[\frac{2}{f(\alpha)}\right]-1} \partial^l \right] \partial_l V_D^i = 0. \tag{3.18}$$

For irrotational motions of the complex system dynamics, the complex velocity fields (3.5) become

$$\hat{V}^i = -2i\lambda(dt)^{\left[\frac{2}{f(\alpha)}\right]-1} \partial^i \ln \Psi, \tag{3.19}$$

where

$$\chi = -2i\lambda(dt)^{\left[\frac{2}{f(\alpha)}\right]-1} \ln \Psi \qquad (3.20)$$

is the complex scalar potential of the complex velocity fields (3.5) and Ψ is the state function (on the significance of Ψ, see Nottale (1993, 2011)). In these conditions, substituting (3.19) in (3.16) and using the mathematical procedures from Agop and Paun (2017), the geodesics equation (3.16) takes the form of the multifractal Schrödinger-type equation

$$2\lambda^2(dt)^{\left[\frac{4}{f(\alpha)}\right]-2} d^l d_l \Psi + i\lambda(dt)^{\left[\frac{2}{f(\alpha)}\right]-1} d_t \Psi. \qquad (3.21)$$

Therefore, for the complex velocity fields (3.19), the dynamics of any complex system are described through multifractal-type Schrödinger "regimens" (i.e. Schrödinger-type equations at various scale resolutions).

Moreover, if Ψ is chosen in the form

$$\Psi = \sqrt{\rho}e^{is}, \qquad (3.22)$$

where $\sqrt{\rho}$ is the amplitude and s is the phase, then the complex velocity fields (3.19) take the explicit form

$$\hat{V}^i = 2\lambda(dt)^{\left[\frac{2}{f(\alpha)}\right]-1} \partial^i s - i\lambda(dt)^{\left[\frac{2}{f(\alpha)}\right]-1} \partial^i \ln \rho, \qquad (3.23)$$

which implies the real velocity fields

$$V_D^i = 2\lambda(dt)^{\left[\frac{2}{f(\alpha)}\right]-1} \partial^i s, \qquad (3.24)$$

$$V_F^i = \lambda(dt)^{\left[\frac{2}{f(\alpha)}\right]-1} \partial^i \ln \rho. \qquad (3.25)$$

By (3.22), (3.24) and (3.25) and using the mathematical procedures from Agop and Păun (2017), Agop and Mercheş (2019), Agop *et al.* (2019), the geodesics equation (3.21) is reduced to the multifractal hydrodynamic-type equations

$$\partial_t V_D^i + V_D^l \partial_l V_D^i = -\partial^i Q, \tag{3.26}$$

$$\partial_t \rho + \partial_l \left(\rho V_D^l \right) = 0, \tag{3.27}$$

with Q the specific multifractal potential

$$Q = -2\lambda^2 (dt)^{\left[\frac{4}{f(\alpha)} \right] - 2} \frac{\partial^l \partial_l \sqrt{\rho}}{\sqrt{\rho}} = -V_F^i V_F^i - \frac{1}{2}\lambda (dt)^{\left[\frac{2}{f(\alpha)} \right] - 1} \partial_l V_F^l. \tag{3.28}$$

Equation (3.26) corresponds to the specific multifractal-type momentum conservation law, while equation (3.27) corresponds to the multifractal-type states density conservation law. The specific multifractal potential (3.28) implies the specific multifractal force

$$F^i = -\partial^i Q = -2\lambda^2 (dt)^{\left[\frac{4}{f(\alpha)} \right] - 2} \partial^i \frac{\partial^l \partial_l \sqrt{\rho}}{\sqrt{\rho}}, \tag{3.29}$$

which is a measure of the multifractality of the motion curves.

Therefore, for the complex velocity fields (3.23), the dynamics of any complex system are described through hydrodynamic "regimens" of a multifractal type (i.e. hydrodynamic equations at various scale resolutions).

The following consequences result:

(i) Any complex system's structural units are in a permanent interaction with a multifractal medium through the specific multifractal force (3.29).

(ii) Any complex system can be identified with a multifractal fluid, the dynamics of which are described by the multifractal hydrodynamic model (see equations (3.26)–(3.28)).

(iii) The velocity field V_F^i does not represent the contemporary dynamics but contributes to the transfer of the specific multifractal-type momentum and to the multifractal-type energy focus. This can be clearly seen from the absence of V_F^i from the multifractal-type states density conservation law and also from its role in the multifractal variational principles (for details, see Mandelbrot (2006)).

(iv) If a multifractal-type tensor is chosen:

$$\hat{\tau}^{il} = 2\lambda^2 (dt)^{\left[\frac{4}{f(\alpha)}\right]-2} \rho \hat{\partial}^i \partial^l \ln \rho, \tag{3.30}$$

the equation defining the multifractal-type "forces" that derive from a multifractal-type "potential" Q can be written in the form of a multifractal-type equilibrium equation. This equation can be written in a tensorial form:

$$\rho \partial^i Q = \partial_l \hat{\tau}^{il}. \tag{3.31}$$

The multifractal-type tensor $\hat{\tau}^{il}$ can be written in the form

$$\hat{\tau}^{il} = \eta \left(\partial_l V_F^i + \partial_i V_F^l \right), \tag{3.32}$$

with

$$\eta = \lambda (dt)^{\left[\frac{2}{f(\alpha)}\right]-1} \rho. \tag{3.33}$$

This is, indeed, a linear constitutive equation of a multifractal type for a multifractal-type "viscous fluid" and allows an original interpretation of the η coefficient as a multifractal-type dynamic viscosity of a multifractal-type fluid.

3.4. Information and multifractal interactions

Let us consider a positive probability density $\rho(x) = \psi(x)\acute{\psi}(x)$ and a finite set of constraints:

$$\int \rho(x) dx = 1, \tag{3.34}$$

$$\int f_k(x) \rho(x) dx = \overline{f_k}, \quad k = 1, \dots, n. \tag{3.35}$$

Now, we can find a probability density $\rho(x)$ which minimizes Shannon's information (Marinescu, 2011)

$$H(\rho, m) = \int \rho(x) \ln \frac{\rho(x)}{m(x)} dx \tag{3.36}$$

bound to conditions (3.34) and (3.35). Note that (3.36) can be invariant to any transformation group through quantity $m(x)$. Usually, the quantity $m(x)$ is identified with the invariant function of the transformation group selected for analysis. Thus, if the group selected for analysis is $SL(2R)$, then the integral invariant function is unitary (Dresselhaus *et al.*, 2010). Then, from a stochastic point of view, it can be stated that the variables of this group are uniformly distributed, a situation in which the principle that minimizes the transformation (3.36) is identified with that of the maximum Onicescu's informational energy $E(\rho)$ (Alipou and Mohajeri, 2012):

$$E(\rho) = \int \rho^2(x)dx. \tag{3.37}$$

In this conjecture, a standard method for finding the minimum of (3.36) is to use the Lagrange multipliers method (Arnol'd, 2010), corresponding to certain given constraints, in order to obtain the functional expression

$$I(\rho, m) = \int \rho(x) ln \frac{\rho(x)}{m(x)} dx + \beta \int \rho(x) dx + \sum_{1}^{m} \lambda_m \int f_k(x) \rho(x) dx. \tag{3.38}$$

By equating to zero, the variation of this functional expression, with respect to $\rho(x)$, the following equation will be obtained:

$$ln \frac{\rho(x)}{m(x)} + 1 + \beta + \sum_{1}^{m} \lambda_k f_k(x) = 0. \tag{3.39}$$

By solving it with respect to $\rho(x)$, we can find

$$\rho(x) = m(x) \exp\left[-\lambda_0 - \sum_{1}^{m} \lambda_k f_k(x)\right], \tag{3.40}$$

where the notation $\lambda_0 = \beta + 1$ has been used.

The minimum variation of information can now be expressed with respect to the multipliers λ_k and the values $\overline{f_k}$ by multiplying (3.39) with $\rho(x)$ and integrating. We obtain

$$H_{\min}(\rho, m) = -\lambda_0 - \sum_{1}^{m} \lambda_k \overline{f_k}. \tag{3.41}$$

Now, it is necessary to choose λ_0 and λ_k in such a way that the constraints (3.34) and (3.35) may be identified.

Under the constraints (3.34)–(3.35) for λ_0, the following value results:

$$\lambda_0 = ln \int m(x) \exp\left[-\sum_1^m \lambda_k f_k(x) \right] dx, \qquad (3.42)$$

which implies the dependence

$$\lambda_0 = lnF\left(\lambda_1, \ldots, \lambda_m\right). \qquad (3.43)$$

In order to obtain under a finite form the function $F\left(\lambda_1, \ldots, \lambda_m\right)$ in the cases where the integration from (3.42) can be carried out, the values $\lambda_1, \ldots, \lambda_m$ can be found through the equation

$$\frac{-\partial}{\partial \lambda_k} \lambda_0 = \overline{f_k}, \qquad (3.44)$$

which results from (3.42) and (3.35).

Unfortunately, it is usually impossible to solve equations (3.42) and (3.44) in a finite form in order to explicitly provide λ_k. There are, however, cases in which these operations can be carried out, and these are the ones in which the *a priori* density is exponential. Indeed, if $m(x)$ is a multivariant exponential,

$$m(x) = \frac{1}{a_1 \cdots a_n} \exp\left[-\sum \frac{x_k}{a_k} \right], \qquad (3.45)$$

where $x = (x_1, \ldots, x_n)$ with x_k real and positive, and if the constraints are

$$\int x_k \rho(x) dx = \overline{x_k}, \quad k = 1, \ldots, n, \qquad (3.46)$$

then the system (3.44) can be immediately solved, which leads to the *a posteriori* density

$$\rho(x) = \frac{1}{\overline{x_1} \cdots \overline{x_n}} \exp\left[-\sum \frac{x_k}{\acute{x}_k} \right]. \qquad (3.47)$$

Thus, the density remains exponential, but with parameters a_k being substituted with the averages extracted from the constraints (3.46).

This is a "convenient" example in which, depending on the nature of the constraints, the *a priori* and *posteriori* densities are the same.

But the exponential can be obtained as an *a posteriori* distribution when it is known that the *a priori* distribution is uniform. Let us look at the case of a single variable, uniformly distributed on any interval of the positive half-line segment.

Imposing the restrictions

$$\int \rho(x)dx = 1 \qquad (3.48)$$

$$\int x\rho(x)dx = \bar{x} \qquad (3.49)$$

for minimizing the function

$$H(\rho, 1) = \int \rho(x)\ln \rho(x)dx, \qquad (3.50)$$

the exponential is obtained:

$$\rho(x) = \frac{1}{\bar{x}}\exp\left(\frac{-x}{\bar{x}}\right). \qquad (3.51)$$

This can be, for example, the case of density $\rho(x)$ for which x is the radial distance r, i.e. $x \equiv r$, and $\bar{x}$ is the average r_0 of the average radial distance. Then, through (3.51), it results in

$$\rho(r) = \frac{1}{r_0}\exp\left(\frac{-r}{r_0}\right). \qquad (3.52)$$

Now, in the case of the radial symmetry r, we can obtain, by substituting this result in the expression of the specific multifractal potential (3.28), the following:

$$Q(r) = -2\lambda^2(dt)^{\left[\frac{4}{f(\alpha)}\right]-2}\frac{1}{\sqrt{\rho}}\left(\frac{d^2\sqrt{\rho}}{dr^2} + \frac{2}{r}\frac{d\sqrt{\rho}}{dr}\right)$$

$$= -2\lambda^2(dt)^{\left[\frac{4}{f(\alpha)}\right]-2}\left(\frac{1}{4r_0^2} - \frac{1}{r_0 r}\right) \qquad (3.53)$$

and also the multifractal specific force

$$F(r) = \frac{-2\lambda^2(dt)^{\left[\frac{4}{f(\alpha)}\right]-2}}{r_0 r^2}. \qquad (3.54)$$

3.5. Multifractal dynamics of a Newtonian type

In what follows, we present a few of the characteristics of the motion induced by the multifractal force field (3.54) using the method described by Mazilu *et al.* (2020). By means of this, the results from the above-mentioned reference are expanded upon for any scale resolution (be it atomic, infra-galactic, extra-galactic, etc.).

In the case where the inertial effects of a multifractal type $\partial_t V_D^i$ are dominant with respect to the convective effects of a multifractal type $V_D^l \partial_l V_D^i$, meaning that the condition $\left|\partial_t V_D^i\right| \gg \left|V_D^l \partial_l V_D^i\right|$ is satisfied, the specific multifractal-type momentum conservation law (3.18), with the constraint (3.54), is given, in a vectorial notation, by the equation

$$\ddot{\boldsymbol{r}} + \frac{k}{r^3}\boldsymbol{r} = 0, \tag{3.55}$$

with

$$k = \frac{2\lambda^2 (dt)^{\left[\frac{4}{f(\alpha)}\right]-2}}{r_0}. \tag{3.56}$$

Here, k is a physical constant that depends both on the scale resolution given by $\lambda^2 (dt)^{\left[\frac{4}{f(\alpha)}\right]-2}$ as well as on the shielding length r_0, $\boldsymbol{r}$ is the position vector and $\ddot{\boldsymbol{r}}$ is the acceleration vector.

A vectorial multiplication of (3.55) with $\boldsymbol{r}$ has, as a consequence, the essential characteristic of force (3.54): It generates only planar motions (the planar motion is kept):

$$\frac{d}{dt}\left(\boldsymbol{r} \times \dot{\boldsymbol{r}}\right) = \boldsymbol{r} \times \ddot{\boldsymbol{r}} = 0. \tag{3.57}$$

Because of the existence of a plane of motion, it is possible to simplify the geometry of the problem by limiting it to the said plane, in which the coordinates of an "entity" in motion are ξ and η.

Then, (3.55) becomes equivalent to the system

$$\ddot{\xi} + \frac{k}{r^2}\cos\Phi = 0, \quad \ddot{\eta} + \frac{k}{r^2}\sin\Phi = 0. \tag{3.58}$$

Here, r and Φ are the polar coordinates in the plane of motion relative to the attraction center. The magnitude of the velocity, which is involved in the variation of the area "swept" by r, is of the form

$$\dot{a} = \xi \dot{\eta} - \eta \dot{\xi} = r^2 \dot{\Phi}. \tag{3.59}$$

This constant of motion allows the integration of system (3.58) with the analytical form of the trajectory as a direct result. First, the complex variable is defined:

$$X_3 = \xi + i\eta = re^{i\Phi}. \tag{3.60}$$

Then, (3.58) becomes

$$\ddot{X}_3 + \frac{k}{r^2}e^{i\Phi} = 0. \tag{3.61}$$

Now, (3.59) can be used in order to eliminate r^2, which implies

$$\ddot{X}_3 + \frac{k}{\dot{a}}e^{i\Phi}\dot{\Phi} = 0 \therefore \dot{X}_3 = i\left(\frac{k}{\dot{a}}e^{i\Phi} + w\right), \tag{3.62}$$

where

$$w = w_1 + iw_2, \quad i = \sqrt{-1} \tag{3.63}$$

is a complex constant which is obviously determined through initial conditions. The analytic equation of motion is obtained from (3.59) and the second equation (3.62), which in polar coordinates of the plane of motion leads to the expression

$$\frac{\dot{a}}{r} = \frac{k}{\dot{a}} + w_1 \cos\Phi + w_2 \sin\Phi \tag{3.64}$$

or, in (ξ, η) coordinates,

$$\left(\frac{k^2}{\dot{a}^2} - w_1^2\right)\xi^2 - 2w_1 w_2 \xi\eta + \left(\frac{k^2}{\dot{a}^2} - w_2^2\right)\eta^2 + 2\dot{a}\left(w_1\xi + w_2\eta\right) = \dot{a}^2. \tag{3.65}$$

This result is a conic. The center of the conic does not coincide with the center of the force, but rather it has the coordinates given through the initial conditions

$$\xi_0 = \frac{-\dot{a}w_1}{\Delta}, \quad \eta_0 = \frac{-\dot{a}w_2}{\Delta}, \quad \Delta = \left(\frac{k}{\dot{a}}\right)^2 - w_1^2 - w_2^2. \tag{3.66}$$

For $\Delta = 0$, the geometric center of the trajectory is located at infinity and the trajectory becomes a parabola.

Assuming, however, that the center of the trajectory is located at a finite distance and relating the geometric description of the trajectory to said center through the translation

$$X_1 = \xi - \xi_0, \quad X_2 = \eta - \eta_0, \tag{3.67}$$

the trajectory equation becomes

$$\left(\frac{k^2}{\dot{a}^2} - w_1^2\right) X_1^2 - 2w_1 w_2 X_1 X_2 + \left(\frac{k^2}{\dot{a}^2} - w_2^2\right) X_2^2 = \frac{k^2}{\Delta}. \tag{3.68}$$

This is still a conic, with the difference that the relationship is related to its center, and the orientation (i.e. the direction of its center with respect to the center of the multifractal force) is given by the vector (w_1, w_2). In such a situation, the supplementary condition $\Delta > 0$, which shows that the vector (w_1, w_2) is limited in modulus, specifies the fact that the trajectory is not an ellipse anymore, but rather a hyperbola or a parabola.

It is obvious that in (3.65), the issue shifts toward a conic for which the line

$$w_1 \xi + w_2 \eta - \dot{a} = 0 \tag{3.69}$$

is the polar of the multifractal force center. This lends direct kinematic significance to the coordinates of the conic center, which are determined through the initial motion conditions. As such, the following temporal characterization of the trajectory can be given: An entity (i.e. a particle) launched with any initial velocity in the space of a center which exerts a multifractal attraction force inversely proportional to the square of the distance, eventually describes a conic in the plane of motion. The initial velocity of the entity determines the relative position of the center of motion with respect to the center of the multifractal force. If the magnitude of this velocity is below a certain limit, then the trajectory is an ellipse. Otherwise, this trajectory is either a parabola or a hyperbola. Moreover, as it results from the above, the center of the multifractal force is different from the center of the multifractal trajectory, with the measure of this difference being the eccentricity, which depends on the initial conditions. A set of such eccentricities can be found, all of them depending on the initial conditions and reflecting all the "possibilities" of realization of Newtonian-type multifractal dynamics.

3.6. Geometries involved in multifractal dynamics of a Newtonian type

The set of conics can be structured as a Cayleyan space, in which the squared forms from the conics' equations in Cartesian coordinates are represented through points with coordinates given by their coefficients. The absolute of this space is given in the context of form (3.65) of the trajectory equation by the conic

$$\frac{k^2}{\bar{a}^2} - w_1^2 - w_2^2 = 0. \tag{3.70}$$

This conic represents a circle in the plane of velocity motions induced by the multifractal force field (3.54). The points inside the circle represent, in absolute geometry, ellipses and relate to the initial velocities that are upper bounded in magnitude by a certain value dictated by the multifractal force field (3.54). The points on the absolute represent parabolas, and the points outside the absolute represent hyperbolas. Let it be noted that the nature of the trajectory in the multifractal force field (3.54) is dictated by the initial velocity of the entity and that, for the current trajectory to be an ellipse, the initial velocity must be upper bounded in modulus. The geometry of the initial velocities plane is a hyperbolic geometry, and it is a natural geometry of the velocity space in special relativity (Misner *et al.*, 2018).

Thus, the families of trajectories in the multifractal force field (3.54) can be systematically characterized through a non-Euclidean geometry of the relative positions of the motion centers, which are represented by these trajectories with respect to the unique center of the multifractal force field (3.54). First of all, the parabolic trajectories are all represented by points on the circle (3.70), which can be written in the form

$$\mu^2 + \nu^2 = 1, \quad \mu = e\cos\Omega, \quad \nu = e\sin\Omega, \tag{3.71}$$

where e represents the eccentricity of the trajectory, defined by the relations

$$a^2 = \frac{k^2}{\Delta^2}, \quad b^2 = \frac{\acute{a}^2}{\Delta}, \quad e^2 = \frac{a^2 - b^2}{a^2} \left(\frac{\dot{a}}{k}\right)^2 (w_1^2 + w_2^2), \quad (3.72)$$

with a and b the half-axes of the orbit. Thus, the initial condition can actually be expressed only in terms of "contemporary" magnitudes, allowing the discarding of previous considerations:

$$w_1 = \frac{k}{\dot{a}} e \cos \Omega, \quad w_2 = \frac{k}{\dot{a}} e \sin \Omega \quad (3.73)$$

The absolute metric of the interior of the circle (3.71) has the expression

$$ds^2 = \frac{(1 - \nu^2)(d\mu)^2 + 2\mu\nu d\mu d\nu + \left(1 - \mu^2\right)(d\nu)^2}{1 - \mu^2 - \nu^2} \quad (3.74)$$

and can be distilled to a form given by Poincaré for the metric of the hyperbolic plane, or by the one given by Lobachevsky,

$$ds^2 = -4 \frac{dh d\bar{h}}{\left(h - \bar{h}\right)^2} = \frac{-du^2 + dv^2}{v^2}, \quad (3.75)$$

through the following coordinates transformations:

$$\mu = \frac{h\bar{h} - 1}{h\bar{h} + 1}, \quad \nu = \frac{h + \bar{h}}{h\bar{h} + 1}, \quad (3.76)$$

$$h = u + iv = \frac{\nu + i\sqrt{1 - \mu^2 - \nu^2}}{1 - \mu}, \quad \bar{h} = u - iv. \quad (3.77)$$

The metric (3.75) is a differential invariant of a continuous group with three parameters (a group of $SL(2R)$ type) (Agop and Păun, 2017; Agop and Mercheş, 2019) of the complex plane, whose infinitesimal generators are the Killing vectors of the metric.

In short, the projection of the "momentum" forms generated by the metric on these Killing vectors are differential forms, which represent constant rates along the metric geodesics. The three Killing vectors have the expressions (Agop and Păun, 2017; Agop and Mercheş, 2019)

$$\hat{B}_1 = \frac{\partial}{\partial u}, \quad \hat{B}_2 = u\frac{\partial}{\partial u}+v\frac{\partial}{\partial v}, \quad \hat{B}_3 = (u^2 - v^2)\frac{\partial}{\partial u}+2uv\frac{\partial}{\partial v}. \quad (3.78)$$

Now, the two components of the "momentum" vector can be obtained by considering the metric (3.75) as a Lagrangean of the geodesic motion. This provides, for the "momentum", the differential forms

$$p_u = \frac{du}{v^2}, \quad p_v = \frac{dv}{v^2}. \quad (3.79)$$

By projecting this vector on the Killing vectors from (3.78), the differential forms result in

$$\omega_1 = \frac{du}{v^2}, \quad \omega_2 = 2\frac{udu + vdv}{v^2}, \quad \omega_3 = \frac{(u^2 - v^2)\,du + 2uvdv}{v^2}. \quad (3.80)$$

Then, it is easy to check that the metric (3.75) can be reproduced by the squared expression

$$\frac{1}{4}\left(\omega_2^2 - 4\omega_1\omega_3\right) \quad (3.81)$$

and that the differential forms from (3.80) are proportional to the elementary arch of the geodesics of metric (3.75), which are given by the parametric equations

$$u(s) = u_0 + v_0 \tanh(v_0 s), \quad v(s) = \frac{v_0}{\cosh(v_0 s)}, \quad (3.82)$$

with u_0 and v_0 constant.

Now, in terms of parameters (e, Ω), i.e. the eccentricity and orientation of the orbit in its plane, the metric (3.74) becomes

$$ds^2 = \left(\frac{de}{1 - e^2}\right)^2 + \frac{e^2}{(1 - e^2)}(d\Omega)^2. \quad (3.83)$$

Moreover, it is possible to rewrite the metric (3.78) by noticing that, for elliptic trajectories, "e" is confined to the interval between -1

and +1. Thus, through the change of parameter

$$e = \tanh \chi, \tag{3.84}$$

the metric (3.78) takes the form

$$ds^2 = (d\chi)^2 + \sinh^2\chi(d\Omega)^2. \tag{3.85}$$

The parameter h from (3.77) has a direct relationship with the Ernst potential of general relativity (Misner *et al.*, 2018; Jaffe and Taylor, 2018) via a harmonic map.

In order to show this relationship, the parameter h is rewritten in terms of (e, Ω). The expression is

$$h = i\frac{\cosh\theta + \sinh\theta e^{-i\Omega}}{\cosh\theta - \sinh\theta e^{-i\Omega}}, \quad \chi = 2\theta. \tag{3.86}$$

It happens that (3.81) represents a harmonic map for the usual space into the Lobachevsky plane, provided χ (and therefore θ) is a solution to the Laplace equation in free space:

$$\Delta\chi = 0. \tag{3.87}$$

Indeed, the problem of harmonic correspondences between the usual space and the hyperbolic plane is described by the stationary values of the energy functional corresponding to the metric (3.75). This is defined as the volume integral of an integrand obtained from that metric by transforming the differentials into space gradients (Misner *et al.*, 2018; Xi, 2018).

The stationary values of the energy functional therefore correspond to the solutions to the Euler–Lagrange equations for a Lagrangean, namely

$$L = -4\frac{\gamma^{il}\partial_i h \partial_l \overline{h}}{(h - \overline{h})^2}, \tag{3.88}$$

with γ^{il} the metric of the usual space. Then, through

$$\delta \iiint \frac{\gamma^{il}\partial_i h \partial_l \overline{h}}{(h - \overline{h})^2}\sqrt{det(\gamma)}d^3 x, \tag{3.89}$$

it results in

$$(h - \overline{h})\partial^l\partial_l h = 2\partial_l h \partial^l h \tag{3.90}$$

and its complex conjugate. Then, it can be seen that h from (3.86) verifies (3.90) when χ is a solution to the Laplace equation and Ω

is arbitrary (in the sense that it does not depend on the position space). In the case of $h \equiv i\varepsilon$, the variational principle (3.89) in the form (Ernst principle (Misner *et al.* 2018; Jaffe and Taylor, 2018; Xi, 2018))

$$\delta \iiint \frac{\gamma^{il}\partial_i \varepsilon \partial_l \bar{\varepsilon}}{(\varepsilon + \bar{\varepsilon})^2} \sqrt{det(\gamma)} d^3 x = 0 \qquad (3.91)$$

refers exclusively to the complex potential of Ernst (Misner *et al.*, 2018; Jaffe and Taylor, 2018), i.e. to the gravitational field in vacuum. Therefore, if the variational principle (3.89) is accepted in form (3.91) as a starting point, the main purpose of the gravitational field research will be to produce the metrics of the Lobachevsky plane, or the metrics correlated with this plane (because they are closely connected to Einstein's field equations by means of operational procedures described by Agop and Merches (2019), Misner *et al.* (2018), Jaffe and Taylor (2018) and Xi (2018). But such a variational principle exceeds the theoretical frame of employing the Ricci tensor. In fact, one can give examples of its applicability in cases which have nothing to do with Einstein's field equations (Agop and Merches, 2019; Misner *et al.*, 2018).

In what follows, we explain the harmonic mapping modes based on both scale resolution and temporal ordering. In Figure 3.1, we show the 3D representation (left side) and contour plot (right side) for the real $(\text{Re}(h))$ and imaginary $(\text{Im}(h))$ parts of h as well as the full representation of the signal $(\text{Se}(h))$, where $r = \tanh\left(\frac{\chi}{2}\right)$, $\Omega = \omega t$ and ω is the pulsation of motion and t is the time. It is possible to see a multi-structuring and modulation of the signal in all three representations. The real part of h, $(\text{Re}(h))$, presents three structures in both time and ω coordinate, while $\text{Im}(h)$ presents multiple similar structures which are seen to blend together for a short amount of time. When representing the complete function $(\text{Se}(h))$, we can observe that the structure follows the patterns imposed by the implicit contribution on the modulated dynamics, through $\text{Im}(h)$, while the nature of the modulations is given by the explicit contribution through $\text{Re}(h)$.

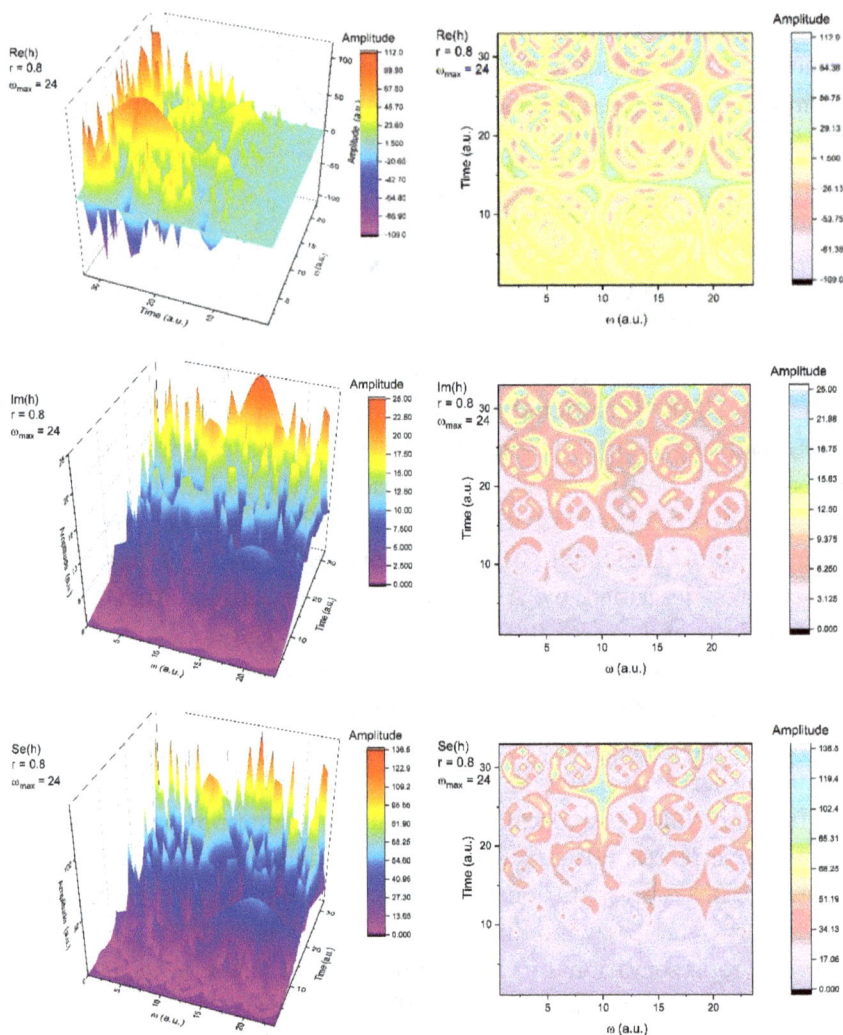

Figure 3.1. Three-dimensional representation and contour plots of the real (Re(h)) and imaginary (Im(h)) parts of h as well as the full representation of the signal (Se(h)).

Thus, we can find three different representations of the function which reflect the explicit (Re(h)), implicit (Im(h)) and measurable (Se(h)) contributions. When looking for the impact of the external factors, through the control parameter r, it is possible to note that various dynamics can be induced at different scale resolutions ($\omega_{\max}$). In Figure 3.2, we have represented the time series of the three aforementioned functions for a fixed control parameter, $r = 0.8$, which will depict a dynamic unperturbed by external factors for two different scale resolutions ($\omega = 24$ shown on the left-hand side and $\omega = 31$ shown on the right-hand side). It can be observed that the period doubling is actually a superposition of the real and imaginary parts of h. The real and imaginary parts depict classical oscillatory behavior with different frequencies. The changes in the scale resolution would automatically be reflected by the shape and structure of the time series (right-hand side of Figure 3.2). One can observe a complicated structure in the imaginary plane, where a modulated structure overlapping an intermittent dynamic can be seen, while in the real plane, a high-frequency structure can be observed. These fascinating results come from the representation in the measurable plane, which is seen here as quasi-chaotic and mostly following the structure imposed by $\text{Im}(h)$. This means that for some systems in which the explicit form is not chaotic, nonlinear behaviors can transpire from the implicit information where chaos can be "hidden".

In order to quantify or even showcase the implicit chaos, as suggested above, the attractors for all the time series represented in Figure 3.3 have been reconstructed. For $\omega = 31$, it is possible to observe single-frequency dynamics with the difference being in the "flatness" of the attractor in response to the changes in the frequencies between the real and imaginary parts. The $\text{Se}(h)$ attractor showcases the double period structure. For $\omega = 31$, a modulated behavior for $\text{Re}(h)$ is observed, while $\text{Im}(h)$ presents a chaotic structure with multiple trajectories seen in the projections in the XZ or XY plane. This attractor seems to describe wide trajectories converging to a "focal area". The $\text{Se}(h)$ attractor depicts similar trajectories as in the case of $\text{Im}(h)$, with the main difference being that the higher density of trajectories are at the base of the attractor, becoming scarcer toward the edge of the attractor.

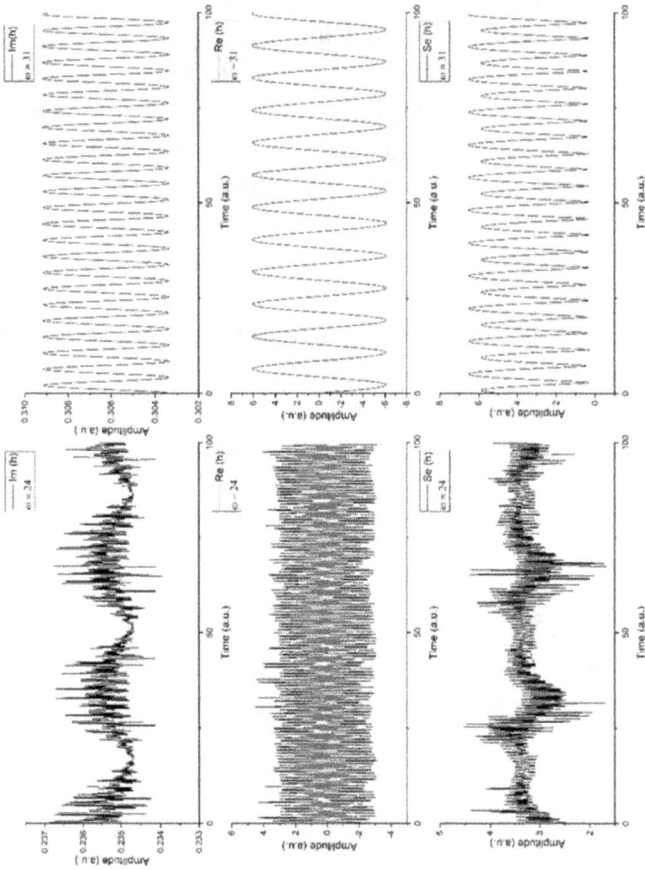

Figure 3.2. Time series of Re(h), Im(h) and Se(h) for two different scale resolutions: (left) $\omega = 24$ and (right) $\omega = 31$.

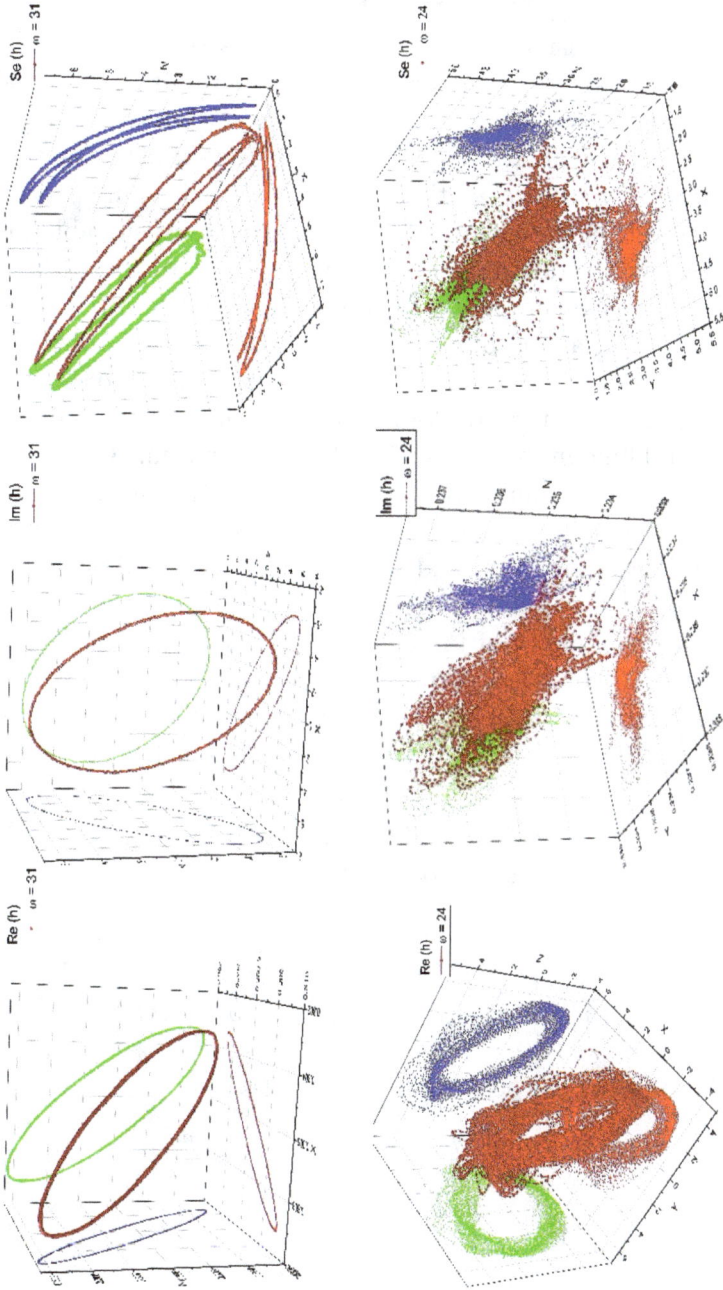

Figure 3.3. Reconstruction of the system attractors in the phase space for Re(h), Im(h), and Se(h) at two resolution scales ($\omega = 24$ and $\omega = 31$).

3.7. Dynamics in laser ablation plasma

The laser ablation plasma, both functionally and structurally, can be considered a multifractal structure. Such a hypothesis is sustained by the following typical example: Between two successive collisions, the trajectory of any ablation plasma particle is a straight line that becomes non-differentiable at the impact point. The fundamental hypothesis of such a statement is that the dynamics of any ablation plasma entity will be described by continuous but non-differentiable multifractal curves. In any case, it is imposed that the explicit form of the velocity field at a non-differentiable scale can be shown through the functionality of evolution equations, which correspond to the canceling of a specific multifractal force and to the incompressibility of the multifractal fluid in the description of ablation plasma. Generally, it is difficult to obtain an analytical solution for this system while taking into account its nonlinear nature induced both by means of non-differentiable convection and by non-differentiable dissipation. However, it is still possible to obtain an analytic solution in the case of a plane symmetry of the dynamics of the non-differentiable ablation plasma. For that purpose, let us consider the equation system of multifractal hydrodynamics at a non-differentiable scale resolution for the stationary case and at plane symmetry as

$$u\partial_x u + v\partial_y u = \sigma \partial_{yy}^2 u, \tag{3.92}$$

$$\partial_x u + \partial_y v = 0, \tag{3.93}$$

where

$$V_F = V_F(u, v), \quad u = u(x, y), \quad v = v(x, y), \quad \sigma = \lambda (dt)^{\left(\frac{2}{f(\alpha)}\right) - 1}. \tag{3.94}$$

Now, by introducing the adimensional variables

$$\xi = \frac{x}{x_0}, \quad \eta = \frac{y}{y_0}, \quad U = \frac{u}{u_0}, \quad V = \frac{v}{v_0}, \tag{3.95}$$

with the property that

$$u_0 y_0 = v_0 x_0, \tag{3.96}$$

The system in equations (3.92) and (3.93) can be rewritten as

$$U\partial_\xi U + V\partial_\eta U = \nu\partial_{\eta\eta}^2 U, \tag{3.97}$$

$$\partial_\xi U + \partial_\eta V = 0, \tag{3.98}$$

where

$$\nu = \frac{\sigma}{\sigma_0} = \frac{\sigma}{y_0 v_0} = \frac{\sigma x_0}{y_0^2 u_0}. \tag{3.99}$$

In the previous equations, x_0 and y_0 are specific lengths, u_0 and v_0 are specific velocities and ν is the fractality degree (a measure of the multifractality of the movement curves of the ablation plasma entities), with all of these quantifying the ablation plasma characteristics at non-differentiable scale resolutions. In general, it is difficult to obtain an analytical solution for the previous equation system while taking their nonlinearity into account. However, for the restrictions

$$\lim_{\eta\to 0} V(\xi, \eta) = 0, \frac{\lim_{\eta\to 0}\partial U}{\partial\eta} = 0, \lim_{\eta\to\infty} U(\xi, \eta) = 0, \quad (3.100)$$

$$q = q_0 \int_{-\infty}^{+\infty} U^2 d\eta = ct., q_0 = \rho\frac{u_0^2}{y_0}, \tag{3.101}$$

the velocity field of the ablation plasma entities at non-differentiable scale resolutions is given by the relations

$$U = \frac{1.5}{(\nu\xi)^{\frac{1}{3}}} sech^2 \left[\frac{0.5\eta}{(\nu\xi)^{\frac{2}{3}}}\right], \tag{3.102}$$

$$V = \frac{1.9}{(\nu\xi)^{\frac{1}{3}}} \left\{\frac{\eta}{(\nu\xi)^{\frac{2}{3}}} sech^2 \left[\frac{0.5\eta}{(\nu\xi)^{\frac{2}{3}}}\right] - \tanh\left[\frac{0.5\eta}{(\nu\xi)^{\frac{2}{3}}}\right]\right\}. \tag{3.103}$$

Thus, the velocity fields of the ablation plasma entities at non-differentiable scale resolutions are composed of solitonic multifractal modes along the $O\xi$ axis respectively solitonic multifractal modes — kink multifractal modes along the $O\eta$ axis (Figures 3.4–3.11).

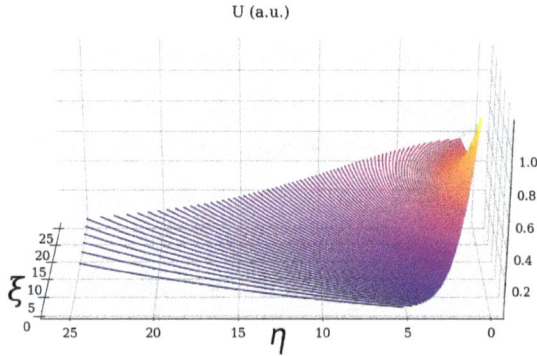

Figure 3.4. Normalized velocity field U of the non-differentiable ablation plasma; $\xi = 0.5$.

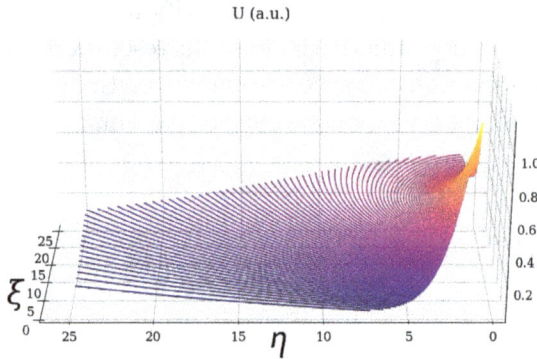

Figure 3.5. Normalized velocity field U of the non-differentiable ablation plasma; $\xi = 1$.

Now, through equations (3.102) and (3.103), the vortex field at non-differentiable scale resolutions is introduced:

$$\Omega = (\partial_\eta U - \partial_\xi V) = \frac{0.57\eta}{(\nu\xi)^2} + \frac{0.63\xi}{(\nu\xi)^{\frac{4}{3}}} \tanh\left[\frac{0.5\eta}{(\nu\xi)^{\frac{2}{3}}}\right]$$

$$+ \frac{1.9\eta}{(\nu\xi)^2} sech^2\left[\frac{0.5\eta}{(\nu\xi)^{\frac{2}{3}}}\right] - \frac{-0.57\eta}{(\nu\xi)^2} \tanh^2\left[\frac{0.5\eta}{(\nu\xi)^{\frac{2}{3}}}\right]$$

$$- \left[\frac{1.5}{\nu\xi} + \frac{1.4\eta}{\xi(\nu\xi)^{\frac{5}{3}}}\right] sech^2\left[\frac{0.5\eta}{(\nu\xi)^{\frac{2}{3}}}\right] \tanh\left[\frac{0.5\eta}{(\nu\xi)^{\frac{2}{3}}}\right].$$

$$(3.104)$$

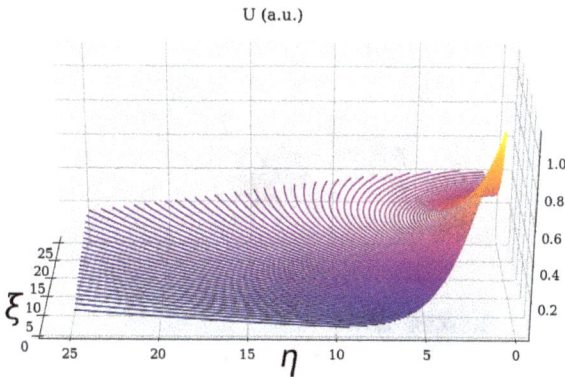

Figure 3.6. Normalized velocity field U of the non-differentiable ablation plasma; $\xi = 1.5$.

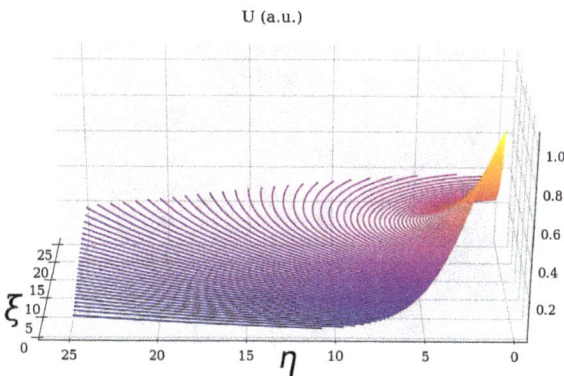

Figure 3.7. Normalized velocity field U of the non-differentiable ablation plasma; $\xi = 2$.

The vortex velocity field and virtual source of turbulence at non-differentiable scale resolutions ares shown in Figures 3.12–3.15.

The minimal vortex field becomes manifest and a real source of turbulence at differentiable scale resolutions through coherence, or the self-structuring of minimal vortices in vortex streets. Indeed, let us consider a two-dimensional, non-differentiable and non-coherent fractal fluid. Then, its entities, assimilated to minimal vortex-type

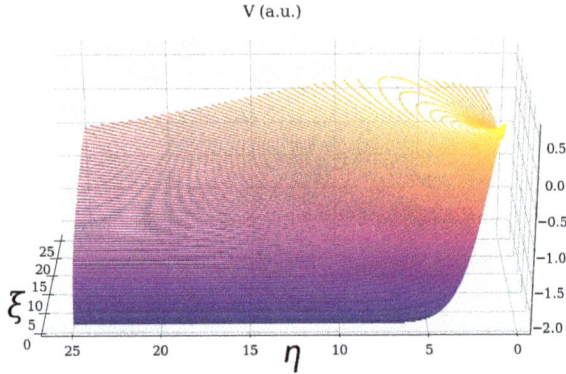

Figure 3.8. Normalized velocity field V of the non-differentiable ablation plasma; $\xi = 0.5$.

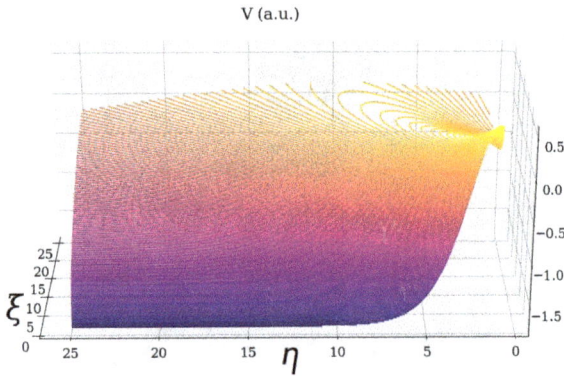

Figure 3.9. Normalized velocity field V of the non-differentiable ablation plasma; $\xi = 1$.

objects, are structured as a two-dimensional vortex lattice of cnoidal stationary modes.

3.8. Free-particle-type problem in the fractal hydrodynamic model

We use equations (3.26) and (3.27) in order to study the dynamics of free perturbation. Due to their nonlinearity, obtaining an analytical solution in the general case can be difficult. However, there are special

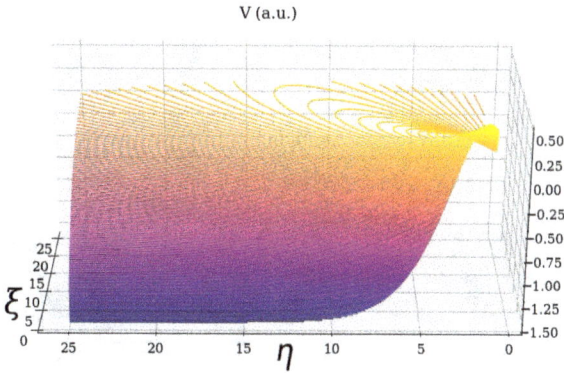

Figure 3.10. Normalized velocity field V of the non-differentiable ablation plasma; $\xi = 1.5$.

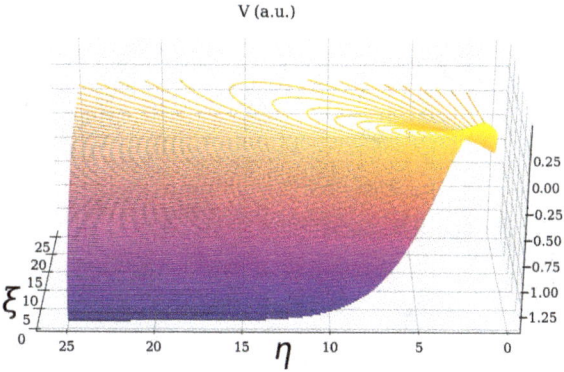

Figure 3.11. Normalized velocity field V of the non-differentiable ablation plasma; $\xi = 2$.

situations in which obtaining a solution is possible. Indeed, let us consider the one-dimensional case of equations (3.26) and (3.27) in the absence of any external constraints, $U = 0$ and $(\alpha)D_F \equiv$, i.e.

$$\partial_i V_D + V_D \partial_x V_D = 2\lambda^2 (dt)^{\left(\frac{4}{D_F}\right) - 2} \rho^{\frac{-1}{2}} \partial_{xx} \rho^{\frac{1}{2}}, \qquad (3.105)$$

$$\partial_t \rho + \partial_x \rho V_D = 0, \qquad (3.106)$$

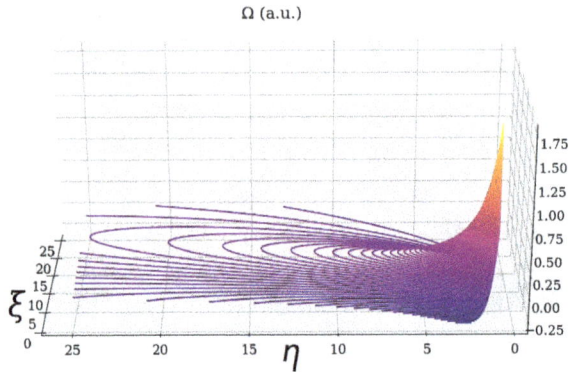

Figure 3.12. Normalized vortex Ω of the non-differentiable ablation plasma; $\xi = 0.5$.

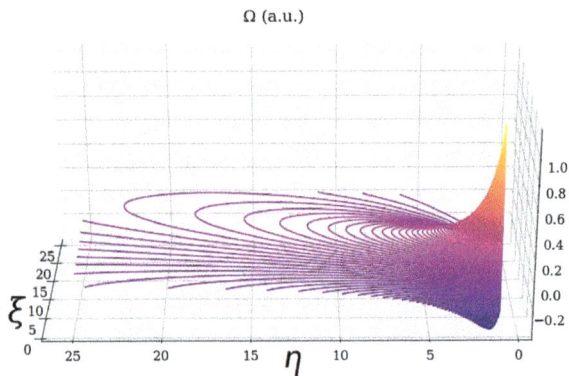

Figure 3.13. Normalized vortex Ω of the non-differentiable ablation plasma; $\xi = 1$.

with the initial conditions

$$V_D(x, t = 0) = c, \tag{3.107}$$

$$\rho(x, t = 0) = \rho_0 \exp\left[-\left(\frac{x}{a}\right)^2\right] \tag{3.108}$$

and the boundary conditions

$$V_D(x = ct, t) = c, \tag{3.109}$$

$$\rho(x = -\infty, t) = \rho(x = \infty, t) = 0. \tag{3.110}$$

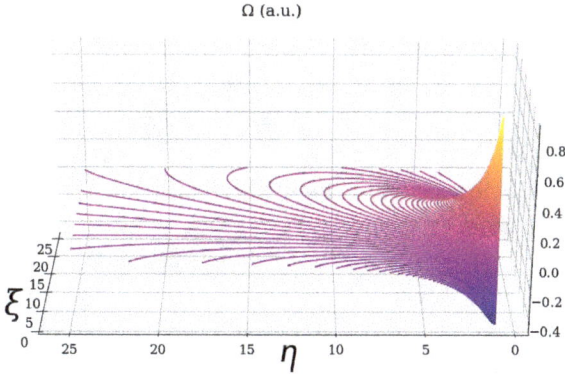

Figure 3.14. Normalized vortex Ω of the non-differentiable ablation plasma; $\xi = 1.5$.

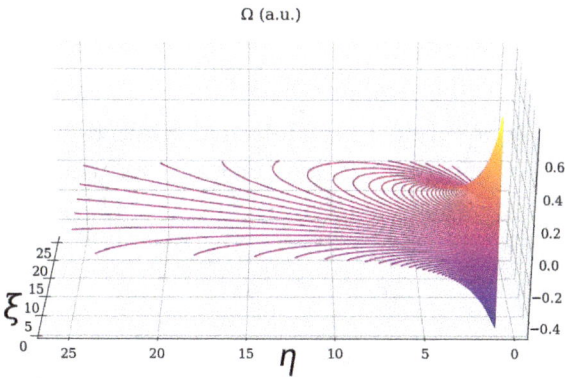

Figure 3.15. Normalized vortex Ω of the non-differentiable ablation plasma; $\xi = 2$.

At any $t > 0$ or $t < 0$, $\partial_x Q \geq 0$ results (for details, see Agop and Păun (2017)). Since $x > ct$ in the absence of external forces, this suggests that equation (3.105) can be separated into

$$\partial_x \left(\rho^{\frac{-1}{2}} \partial_{xx} \rho^{\frac{1}{2}} \right) = \frac{2}{a(t)^2}(x - ct) \tag{3.111}$$

and

$$\partial_t v + v \partial_t v = \left[\frac{2\lambda}{a(t)} \right]^2 (dt)^{\left(\frac{4}{D_F} \right) - 2}(x - ct). \tag{3.112}$$

Integration of equation (3.111) gives, taking the boundary condition from (3.110) into consideration, a quadratic solution in $(x - ct)$, i.e.

$$\rho(x,t) = \frac{1}{[\pi a(t)]^{\frac{1}{2}}} \exp\left[\frac{-(x-ct)^2}{a(t)}\right]. \tag{3.113}$$

Indeed, this function satisfies the initial condition (3.108) if the initial value of $a(t)$ is chosen as

$$a(t = 0) = \alpha^2. \tag{3.114}$$

The insertion of equation (3.113) into the continuity equation (3.106) indicates that, for $x = ct$,

$$(2a)^{-1}d_t a = (\partial_x V_D)_{x=ct}. \tag{3.115}$$

Then, the differential equation for $a(t)$ is obtained by performing the operation $(\partial_x)_{x=ct}$ on equation (3.112):

$$a d_{tt} a - \frac{1}{2}(d_t a)^2 = 8\lambda^2 (dt)^{\left(\frac{4}{D_F}\right)-2}. \tag{3.116}$$

The solution to equation (3.116) with the initial condition (3.114) (satisfying the requirement that $\rho(x,t)$ be real, which is met by the symmetry condition $a(t) = a(-t)$, equation (3.113)) is

$$a(t) = \alpha^2 + \left[\frac{2\lambda(dt)^{\left(\frac{2}{D_F}\right)-1}}{\alpha}(t)\right]^2. \tag{3.117}$$

According to equations (3.113) and (3.117), the states density is Gaussian, which is in a form with a time-dependent distribution parameter $a(t)$ and spreads with the "classical" particle velocity c:

$$\rho(x,t) = \frac{\pi^{\frac{-1}{2}}}{\left\{\alpha^2 + \left[\frac{2\lambda(dt)^{\left(\frac{2}{D_F}\right)-1}}{\alpha}t\right]^2\right\}^{\frac{1}{2}}} \exp\left\{\frac{-(x-ct)^2}{\alpha^2 + \left[\frac{2\lambda(dt)^{\left(\frac{2}{D_F}\right)-1}}{\alpha}t\right]^2}\right\}. \tag{3.118}$$

Similarly, integration of equation (3.105) with the initial condition (3.107) and the boundary condition (3.109) gives the velocity field of the particle at the differentiable scale:

$$V_D = \frac{c\alpha^2 + \left[\frac{2\lambda(dt)^{\left(\frac{2}{D_F}\right)-1}}{\alpha}\right]^2 tx}{\alpha^2 + \left[\frac{2\lambda(dt)^{\left(\frac{2}{D_F}\right)-1}}{\alpha}t\right]^2}. \tag{3.119}$$

Equations (3.118) and (3.119) represent the fractal hydrodynamic solution of the free particle at a differentiable scale.

From this point on, we can define the following dynamic variables:

(i) Current density:

$$j(x,t) = \rho(x,t)V_D(x,t) = \frac{c\alpha^2 + \left[\frac{2\lambda(dt)^{\left(\frac{2}{D_F}\right)-1}}{\alpha}\right]^2 tx}{\left\{\alpha^2 + \left[\frac{2\lambda(dt)^{\left(\frac{2}{D_F}\right)-1}}{\alpha}t\right]^2\right\}^{\frac{3}{2}}}$$

$$\times \exp\left\{\frac{-(x-ct)^2}{\alpha^2 + \left[\frac{2\lambda(dt)^{\left(\frac{2}{D_F}\right)-1}}{\alpha}t\right]^2}\right\}. \tag{3.120}$$

(ii) Specific fractal potential:

$$Q = -2\lambda(dt)^{\left(\frac{4}{D_F}\right)-2} = \left\{\frac{(x-ct)}{\alpha^2 + \left[\frac{2\lambda(dt)^{\left(\frac{2}{D_F}\right)-1}}{\alpha}t\right]^2}\right\}^2. \tag{3.121}$$

(iii) Specific fractal force:

$$F(x,t) = -\partial_x Q = \lambda^2 (dt)^{\left(\frac{4}{D_F}\right)-2} = \frac{(x-ct)}{\left\{\alpha^2 + \left[\frac{2\lambda(dt)^{\left(\frac{2}{D_F}\right)-1}}{\alpha}t\right]^2\right\}^2}.$$

(3.122)

For $t > 0$ or $t < 0$, the dynamic variables defined in equations (3.118)–(3.122) are inhomogeneous in x and t, while for $x = ct$, they become

$$V_D(x = ct, t) \to c,$$ (3.123)

$$\rho(x = ct, t) \to \frac{\pi^{\frac{-1}{2}}}{\left\{\alpha^2 + \left[\frac{2\lambda(dt)^{\left(\frac{2}{D_F}\right)-1}}{\alpha}t\right]^2\right\}^{\frac{1}{2}}},$$ (3.124)

$$j(x = ct, t) \to \frac{\pi^{\frac{-1}{2}}c}{\left\{\alpha^2 + \left[\frac{2\lambda(dt)^{\left(\frac{2}{D_F}\right)-1}}{\alpha}t\right]^2\right\}^2},$$ (3.125)

$$Q(x = ct, t) \to 0,$$ (3.126)

$$F(x = ct, t) \to 0.$$ (3.127)

References

Agop M. and Mercheş I. 2019. *Operational Procedures Describing Physical Systems*. Boca Raton: CRC Press.

Agop M. and Păun V. P. 2017. *On the New Perspectives of Fractal Theory: Fundaments and Applicatio*. Bucharest: Romanian Academy Publishing House.

Agop M., Gavriluţ A., and Rezuş E. 2015a. Implications of Onicescu's informational energy in some fundamental physical models. *Int. J. Mod. Phys B*, 29(07), 1550045.

Agop M., Gavriluţ A., Crumpei G., and Doroftei B. 2014. Informational non-differentiable entropy and uncertainty relations in complex systems. *Entropy*, 16(11), 6042–6058.

Agop M., Gavriluţ A., Păun V., Filipeanu D., Luca F., Grecea C., and Topliceanu L. 2016. Fractal information by means of harmonic mappings and some physical implications. *Entropy*, 18(5), 160.

Agop M., Gavriluţ A., Ştefan G., and Doroftei B. 2015b. Implications of non-differentiable entropy on a space-time manifold. *Entropy*, 17(4), 2184–2197.

Agop M., Gravriluţ A., Eva L., and Crumpei G. 2019. *Towards the Multifractal Brain by Means of the Informational Paradigm: Fundaments and Applications.* Iaşi, Romania: Ars Longa Publishing House.

Alipou M. and Mohajeri A. 2012. Onicescu information energy in terms of Shannon entropy and Fisher information density. *Mol. Phys.*, 110(7), 403–405.

Argyris J., Marin C., and Ciubotariu C. 2002. *Physics of Gravitation and the Universe.* Iaşi, Romania: SpiruHaret; Chisinau: Tehnica-Info.

Arnol'd V. I. 2010. *Mathematical Methods of Classical Mechanics.* New York: Springer.

Cresson J. 2003. Scale Calculus and the Schrödinger Equation. *J. Math. Phys.*, 44(11), 4907–4938.

Cristescu C. P. 2008. *Non-linear Dynamics and Chaos: Theoretical Fundaments and Applications.* Bucharest: Romanian Academy Publishing House.

Dresselhaus M. S., Dresselhaus G., and Jorio A. 2010. *Group Theory: Application to the Physics of Condensed Matter.* New York: Springer.

Gavriluţ A., Mercheş I., and Agop M. 2019. *Atomicity Through Fractal Measure Theory: Mathematical and Physical Fundamentals with Applications.* Cham, Switzerland: Springer.

Green S. H. 2000. *Information Theory and Quantum Physics: Physical Foundations for Understanding the Conscious Process.* Berlin: Springer.

Grigorovici A., Bacaiţă E., Păun V., Grecea C., Butuc I., Agop M., and Popa O. 2017. Pairs generating as a consequence of the fractal entropy: Theory and applications. *Entropy*, 19(3), 128.

Hawking, S. W. and Penrose, R. 1996. *The Nature of Space and Time.* Princeton, NJ: Princeton Univ. Press.

Jaffe R. L. and Taylor W. 2018 *The Physics of Energy*, Cambridge University Press.

Lakshmanan M. and Rajaseekar S. 2003. *Nonlinear Dynamics Integrability, Chaos and Patterns*, Cambridge, Berlin: Springer.

Mandelbrot, B. 2006. *The fractal geometry of nature.* New York: W.H. Freeman And Company.

Marinescu D. C. 2011. From classical to quantum information theory. London: Academic.

Mazilu N., Agop M., and Mercheş I. 2020. *Mathematical Principles of Scale Relativity Physics. Concept of Interpretation.* Boca Raton: CRC Press Taylor and Francis Group.

Misner C. W., Thorne K. S., and Wheeler J. A. 2018. *Gravitation.* W. H. Freeman.

Mitchell M. 2011. *Complexity: A Guided Tour.* New York; Oxford: Oxford University Press.

Nottale L. 1993. *Fractal Space-time and Microphysics: Towards a Theory of Scale Relativity.* Singapore; River Edge, NJ: World Scientific.

Nottale L. 2011. *Scale Relativity and Fractal Space-time: A New Approach to Unifying Relativity and Quantum Mechanics.* London: Imperial College Press.

Penrose, R. 2007. *The Road to Reality: A Complete Guide to the Laws of the Universe.* New York: Vintage Books, Cop.

Politi A. and Badii R. 2003. *Complexity: Hierarchical Structures and Scaling in Physics.* Cambridge: Cambridge University Press.

Weinberg S. 1994. *Dreams of a Final Theory.* New York: Vintage Books.

Xi Y. 2018. *Geometry of Harmonic Maps.* New York: Springer.

Chapter 4

Charged Particle Dynamics in Laser-Produced Plasmas

4.1. Peculiar effects in nanosecond laser produced plasmas during the pulsed laser deposition process

Although the fundamental studies presented earlier on pure metallic targets offered a great amount of information on the structure and dynamics of the laser-produced plasma plumes, they cover just one side of the pulsed laser deposition (PLD) problem. The investigations for nanosecond, picosecond and femtosecond ablation regimes were made on the so-called "free flow" regime. During expansion, the plume does not interact with any object (substrates (Canulescu et al., 2009; Chen et al., 2014; González-posada, 2016; Focsa et al., 2017), planar Langmuir probes), and the background pressure is low enough so that there are no supplementary collisions between the ejected particles and the residual gas particles (Geohegan and Puretzky, 1996; Harilal et al., 2014; Wen et al., 2007; Amoruso et al., 2010; Glavin et al., 2015; Schou, 2009). However, when we try to apply the obtained results to the deposition process, we need to take into account the fact that in a typical PLD configuration, the plume expansion is limited by the substrate, and usually, these experiments require higher background pressures. In this section, we present the results of a preliminary study based on ICCD fast camera imaging and LP techniques, focused on the global plume dynamics of an aluminum plasma in PLD conditions. Also, we discuss some peculiar effects related to the LPP interaction with the substrate and the ion dynamics, as measured by the Langmuir probe method.

PLD has received a lot of attention in the past 10 years as one of the best techniques to produce complex films with relatively complicated stoichiometry (Dijkkamp *et al.*, 1987; Bulai *et al.*, 2019; D. Craciun *et al.*, 2014). The technique has proven flexibility in terms of the deposition geometry (Yang *et al.*, 2014) and target or background gas nature (Craciun and Craciun, 1992; Dorcioman *et al.*, 2014). Great advancements have been made toward understanding the fundamental aspects of laser ablation for better control of the thin-film deposition technique. Several diagnostic techniques, such as optical emission spectroscopy (Irimiciuc *et al.*, 2017, 2018; Aragón and Aguilera, 2008), mass spectrometry (Bruzzese *et al.*, 1996) or Langmuir probe (LP) (Irimiciuc *et al.*, 2017, 2014; Kumari *et al.*, 2012; Chen *et al.*, 2014), have been implemented to highlight phenomena such as: plasma structuring (Schou *et al.*, 2004; Irimiciuc *et al.*, 2018; Harilal *et al.*, 2003), molecule formation (Harilal *et al.*, 1996; Vivien *et al.*, 1999), elemental distribution in complex plasmas (Irimiciuc *et al.*, 2019) and ejected particle behavior in various conditions. Out of all the techniques presented in the literature, LP has shown great versatility, being implemented for a wide range of materials (Thestrup *et al.*, 2002; Irimiciuc *et al.*, 2017) and various irradiation conditions (Harilal *et al.*, 2003; Bruzzese *et al.*, 1996). The LP method presents itself as a relatively simple diagnostic tool for plasma investigations, consisting of submerging a metallic electrode (of cylindrical, planar or spherical geometry) in the plasma to collect the ionic and electronic currents or a mixture of the two currents, depending on the probe biasing. The method was theoretically developed and experimentally applied mainly to steady-state discharge plasmas at quasi-thermodynamic equilibrium. Laser-produced plasmas have transient and periodic characters, with all the plasma parameters presenting a complex spatial distribution and temporal dependence and being generally strongly directional. Therefore, the LP theory was adapted by time-sampling the ionic and electronic currents' temporal traces and considering not only the thermal movement of the particles but also their "flow" velocity.

LP techniques offer a good insight into the ejected particle dynamics, relating to a small volume of plasma, and thus can be implemented for axial and angular measurements, and they can easily be adapted for a wide range of diagnostic geometries. In the time-resolved approach to LP diagnostics, it can offer information from a wide set of plasma properties (electron temperature, plasma potential, charged particle density, collision frequency, etc.) during the plasma expansion.

The flexibility of the technique allowed for some exciting findings. Multi-ionic-peak structures were evidenced with the use of a single cylindrical probe (Nica *et al.*, 2010), heated probe (Irimiciuc *et al.*, 2014) and multiple-probe (Tang *et al.*, 2012; Singh *et al.*, 2014) configurations. Their nature is often debated, with some reports presenting their roots in the dynamics of a plasma structure generated by electrostatic mechanisms (Bulgakov and Bulgakova, 1999) or induced by the transient double layers generated through plasma structuring (Focsa *et al.*, 2017). Other novel theoretical approaches have been presented either in the framework of a fractal theoretical model or built around Lorenz-type systems (Irimiciuc *et al.*, 2019). The complete behavior of plasma charge particles can be accounted for in the framework of the aforementioned fractal paradigm, with the multi-peak dynamic being explained through the presence of dissipative or dispersive effects, and the classical behavior being better showcased in the compact fractal hydrodynamic model. The fractal paradigm proposes that any dynamic variable describing the laser ablation plasma systems acts as the limit of a function family. These functions are nondifferentiable for a null resolution scale and differentiable for a nonzero resolution scale. This method is well adapted for applications in the field of laser-produced plasma dynamics, where any real determination conducted at a finite resolution scale clearly implies the development both of a new geometric structure and of a physical theory. The motion laws, invariant to spatial and temporal coordinate transformations, are integrated with scale laws, invariant at scale transformation. Moreover, the motion curves have a dual part: geodesics of the fractal space and streamlines of a fractal fluid, whose entities (ejected particles from the target as a result of laser target interaction) are substituted by their geodesics so that any external constraints are interpreted as a selection of geodesics by means of the diagnostic technique, the LP method.

4.2. Charged particle dynamics in transient plasmas generated by nanosecond laser ablation on Mg target

4.2.1. *Langmuir probe measurements*

For a bias of $+/-5$V applied to the LP, we are able to collect the saturation ionic and electronic currents, as displayed in Figure 4.1(a) and (b) for two different fluences characterized by 2.45 mm^3 plasma volume at 2 cm with respect to the target surface. By recording the saturation charge currents, we are ensuring the collection of the global ionic and electronic charge ejected as a result of the ablation process coupled with the ionization and neutralization processes occurring during the expansion of the laser-produced plasmas. We can distinguish different features for both saturation currents. The electronic current has a longer lifetime, induced by the increased collision rate, and a higher diffusion rate of the electrons as opposed to the ions. For the same applied voltage, the ionic Mg species has a life time of 3 μs, while the electrons present almost twice as much, approximately 6 μs. There is, however, a recurring feature that can be found in all collected signals, regardless of the applied voltage. The choice of representing here only the temporal trace for the saturation currents is supported by their higher amplitude and better highlight of the smaller features, such as the multi-peak regime of the signal noticeable below 1 μs. In the inset of Figure 4.1(a) and (b), we can see a *zoomed-in* view of this regime. We observe that for higher fluences, the first maximum is reached at a shorter expansion time with no significant increase in the amplitude of the current. Results indicate an increase in the kinetic energy of the ejected particles, induced by the increase in laser fluence, while the overall ejected charge remains quasi-constant. This dynamic has been previously

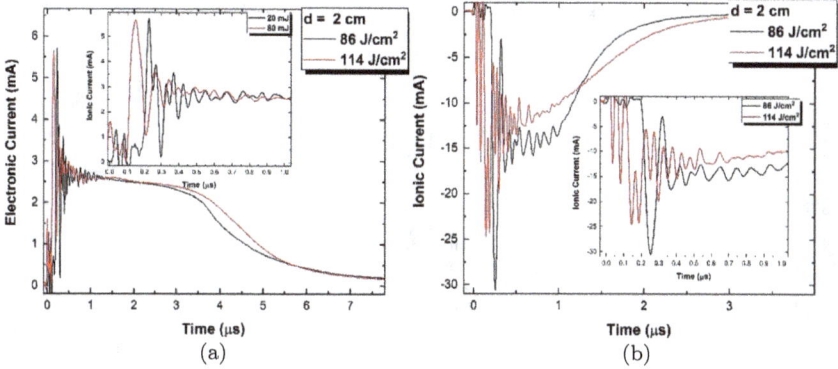

Figure 4.1. Temporal trances of the electronic and ionic currents collected at 2 cm from the target.

reported for a wide range of materials, including classical or more complex ablation geometries (Irimiciuc *et al.*, 2014; Nica *et al.*, 2010; Tang *et al.*, 2012; Singh *et al.*, 2014).

The nature of these features is accepted, in some differentiable theoretical models, to be induced by the transient double layer forming as a result of the electrostatic ejection mechanism. By performing fast Fourier transform (FFT) analysis on the collected signals, we observed two distinct frequencies. These values showcased a strong dependence on the laser fluence (Figure 4.2(a) and (b)) and measurements distance (Figure 4.2(c)). The first frequency is of the order of tens of MHz, ranging from 15 to 20 MHz), while the second one is of the order of a few MHz (ranging from 1 to 10 MHz). The results are in good agreement with those from our previous reports on Ni LPP, where we found values ranging from 5 to 20 MHz, or with those of a wider range of metals reported by Focsa *et al.* (2017). The second frequency is induced by the splitting of the laser-produced plasma during expansion into two plasma structures expanding at different velocities and the appearance of a secondary transient double layer, which will accelerate the slower plasma structure (Marine *et al.*, 2004). Each plasma structure is characterized by a unique frequency. With the increase in the laser

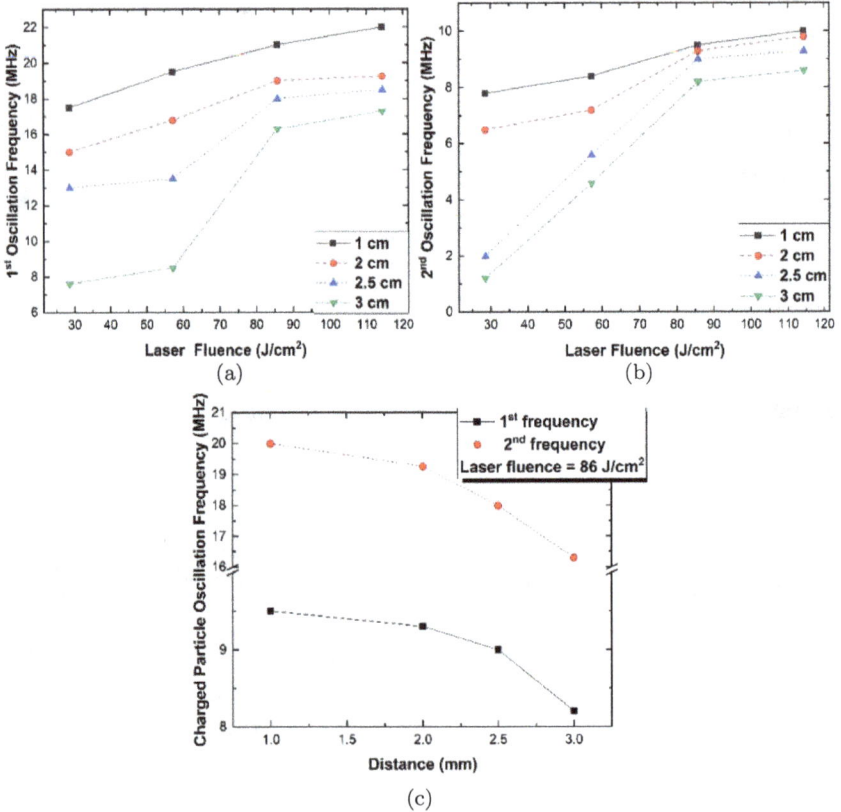

Figure 4.2. Frequency evolution with laser fluence (a) and (b) and measurement distance (c).

fluence, we noticed an increase in the frequency (5–50% depending on the measurement distance), which indicates a generation of a stronger electrical field through Coulomb explosion, followed by saturation for fluences higher than 80 J/cm². The observed features are damped after approximately 1 μs. This can also be seen from the frequency evolution with the measurement distance, where we noticed a decrease of about 25% for both measured frequencies. Admittedly, the laser fluence values are much higher than the ones used in PLD or material processing applications, where we can find reports of fluences below 5 J/cm² (Dijkkamp *et al.*, 1987; Bulai *et al.*,

2019; Craciun *et al.*, 2014; Craciun and Craciun, 1992; Yanh *et al.*, 2014), depending on the irradiation conditions and the nature of the thin film envisioned in each report. However, the results become relevant for the fundamentals of a new generation of high-power laser — matter interactions.

The LP theory is reportedly (Merlino, 2007) limited to only longer expansion times where the theoretical assumptions are met. This allows us to characterize the temporal evolution of some plasma parameters, such as electron temperature, plasma potential or ionic density. In order to achieve that, we implemented the methodology used by Irimiciuc (2017, 2014). Briefly, consecutive bias potentials (in 50 points), ranging from −5 to +10 V, were applied to the heated probe, and the plasma currents were collected from a well-defined plasma volume. The probe was moved to locations at various distances with respect to the target's surface. An example is shown in Figure 4.3. As anticipated from the previous paragraphs, all signals follow similar patterns, with an modulated part for short evolution times followed by a classical decreasing trend.

For a time span of 6 μs and using a step of 1 μs, we reconstruct the *I–V* characteristics for each moment of time. In Figure 4.4, we have presented representative characteristics describing the LPP at various distances after 1 μs (Figure 4.4(a)) and, for a fixed distance of 2.5 cm, the reconstructed characteristics at different moments of time (Figure 4.4(b)). We notice that for an instant temporal sequence, the axial dependence of the *I–V* characteristics is not significant; however, there is a substantial shift toward the positive floating potential. This is due to the spatial distribution of the two sets of charges during expansion. The electrons generally have a more uniform distribution in the plasma; however, in light of the electrostatic ejection mechanism, the first electrons ejected will occupy spatially the position in front of the ejected ions. Although during expansion, there are other phenomena that need to be considered, such as ion neutralization, secondary ionizations or molecule formation (Vivien *et al.*, 1999). This shift in the floating potential showcases the change in ion-to-electron ratio within the plasma volume for a fixed moment in time (1 μs). The spatiotemporal

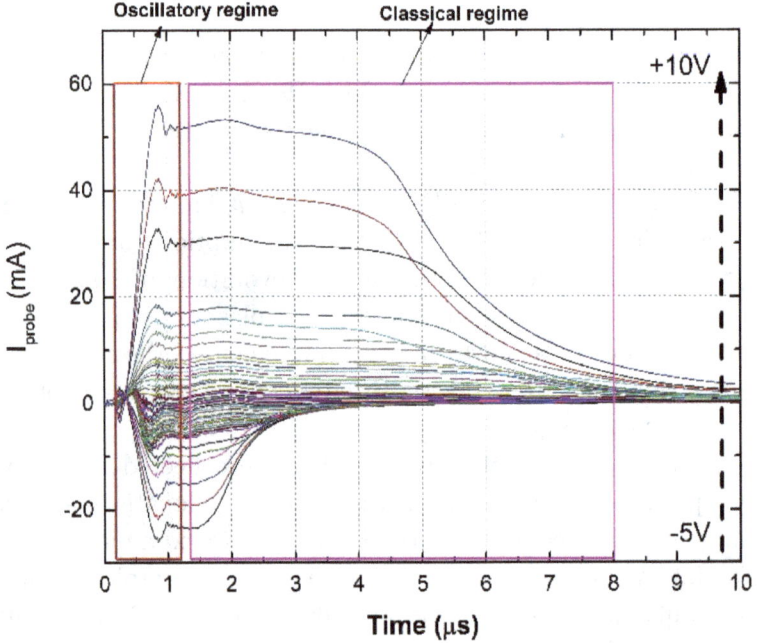

Figure 4.3. Temporal traces of the ionic and electronic currents recorded for a laser fluence of 56 J/cm^2 at 2.5 cm from the target.

evolution of all main plasma parameters considered here (electron temperature and plasma potential) follow a classic quasi-exponential decrease. To determine the aforementioned plasma parameters, we treated the I–V characteristics following the procedure described by Irimiciuc (2017, 2014). Briefly, by applying a logarithm procedure to the I–V characteristic curve, we obtain a distribution which is described by a linear increase, an inflection point and a saturation region. The slope of that linear increase will define the electron temperature, while the inflection point will be the plasma potential (Figure 4.4(e)).

An increase in laser fluence leads to higher electron temperatures and plasma potentials (Figure 4.4(c) and (d)). The highest values found here are $T_e = 1$ eV and $V_p = 9.3$ V for the measurements performed at 1.5 cm and a laser fluence of approximately 114 J/cm^2, while the lowest are found at 28 J/cm^2, with the values decreasing

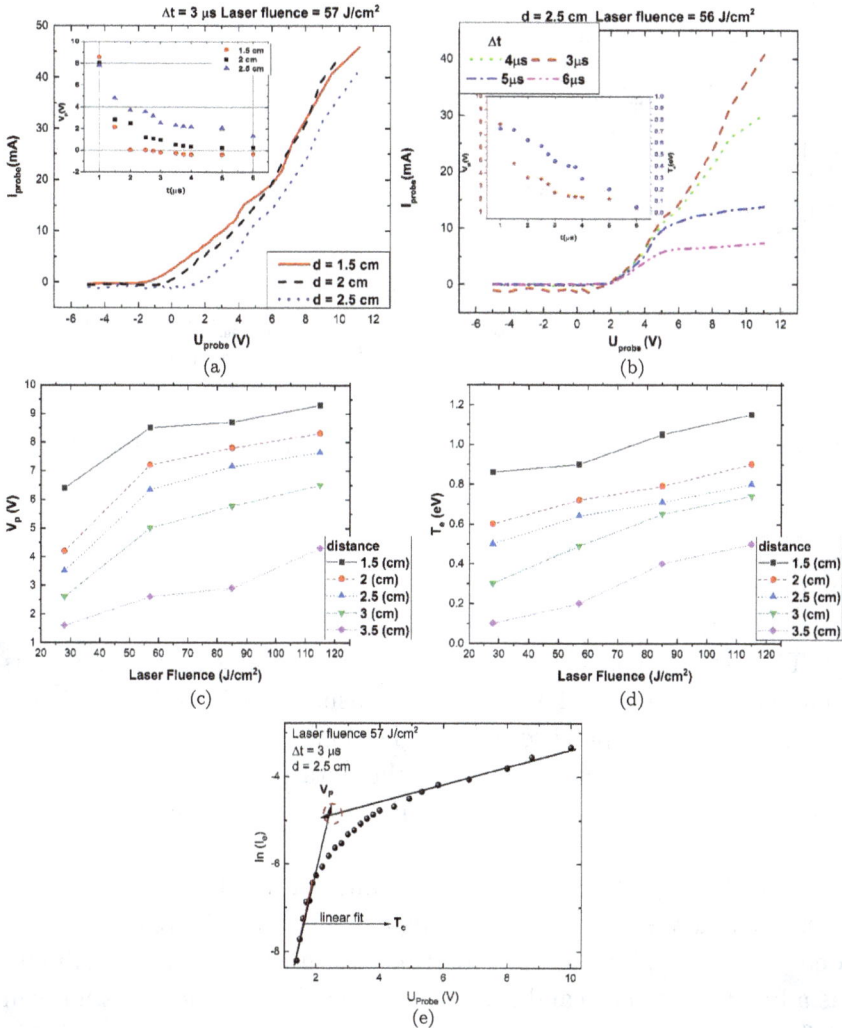

Figure 4.4. Evolution of the I–V characteristics: time (a) and space (b); plasma potential (c) and electron temperature (d) evolution with distance and laser fluence and an example of the logarithmic representation of the I–V characteristics.

by almost one order of magnitude, having $T_e = 0.1$ eV and $V_p = 2$ V. For a fixed distance, the time-resolved analysis reveals important changes in both the shape of the $I - V$ characteristics. These changes induced by decreases in all the plasma parameters are indicators of

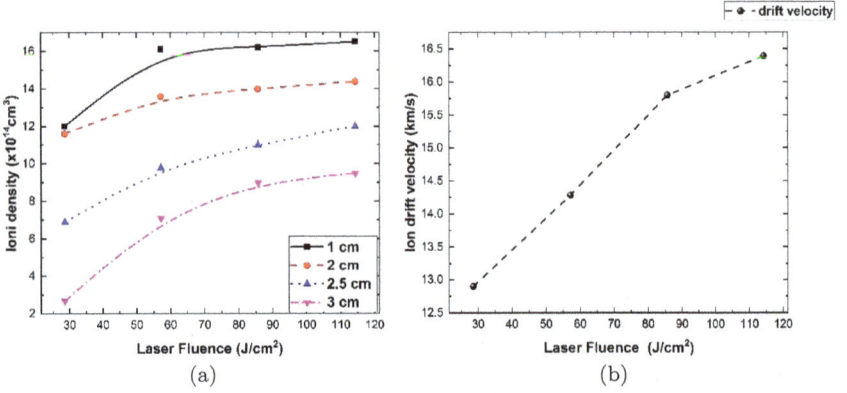

Figure 4.5. Ionic density (a) and velocity (b) evolution with the laser fluence.

the laser-produced plasma expansion and losses in particle density and particle energy, as shown in the inset of Figure 4.4(b), where the temporal evolution of T_e and V_p are presented.

The influence of the laser fluence over some plasma parameters is generally known. A higher laser fluence usually leads to the ejection of a higher density of particles with higher kinetic and thermal energies. For our conditions, we synthesized the data in Figure 4.5, where we present the evolution of the ionic density with laser fluence at various distances (Figure 4.5(a)) and the evolution of the ion drift velocity (Figure 4.5(b)). The results are in line with other reports of an increase followed by a saturation regime. The ion drift velocities were determined by plotting the evolution of the ionic current maximum as a function of space and time. The slope of that representation will define the drift velocity.

The overall effects of the laser fluence depicted in Figures 4.2 and 4.5 are to enhance the modulated charge particle dynamics, increase the current densities and increase the overall kinetic movement of the plasma. The kinetic enhancement of the LPP, however, reaches a saturation regime, where neither the particle density nor the expansion velocities increase, meaning that energy is lost in expelling large structures (clusters and nanoparticles). This result is in line with our previously reported results (Irimiciuc *et al.*, 2014), where the presence of a third plasma structure was seen in the particle velocity distribution at high laser fluence. The interaction between the two plasma structures and their unique signatures in the ionic and electronic saturation currents is yet to be understood. The subtle difference in plasma plume dynamics at high fluence as opposed to the usual lower values used in applications such as PLD still needs to be investigated by other experimental techniques and clarified through comprehensive theoretical modeling.

4.3. Understanding charged particle dynamics in a multifractal paradigm

4.3.1. *Scale covariant derivative and geodesics equations*

The ablation plasma is a fractal medium induced by the collision processes of its entities. Indeed, the laser ablation plasma can be assimilated with a fractal. This assumption can be sustained by a typical example related to the collision processes in the laser ablation plasma: Between two successive collisions, the trajectory of the plasma particle is a straight line that becomes nondifferentiable at the impact point. Considering that all the collision impact points form an uncountable set of points, it results in that the trajectories of plasma particles become continuous but nondifferentiable curves, i.e. a fractal.

In this context, fractal theories of motion become functional for describing the dynamics of ablation plasma. The fundamental assumption of these models is that the dynamics of any entity of ablation plasma will be described by continuous but nondifferentiable motion curves (fractal motion curves). These fractal motion curves exhibit the property of self-similarity at every point, which can be translated into a property of holography (every part reflects the whole). Basically, we are discussing the "holographic implementations" of the dynamics of any ablation plasma entity through Schrödinger-type fractal "regimes" (i.e. describing dynamics by using Schrödinger-type equations at various scale resolutions — Schrödinger equation of fractal type).

4.3.2. *Ablation plasma behavior through a special tunneling effect of fractal type*

In the following, we apply the previously mentioned mathematical model by analyzing the one-dimensional (1D) stationary dynamics of physical systems with spontaneous symmetry breaking in the form of a tunneling effect of fractal type. The results will be correlated or calibrated with the dynamics of a LPP from the perspective of generating two plasma structures as well as from the perspective of some essential characteristics of these structures.

Let us consider the Schrödinger equation of fractal type from Chapter 3, which has an external restriction in the form of the scalar potential U. For the moment, the scalar potential is not explicitly defined.

In the 1D stationary case, this equation becomes

$$\lambda^2 (dt)^{(4/D_F)-2} \partial_{zz} \Psi(z,t) + i\lambda (dt)^{(2/D_F)-1} \partial_t \Psi(z,t) - \frac{U}{2} \Psi(z,t) = 0.$$

(4.1)

If the scalar potential U is time independent, $\partial_t U = 0$, then (4.1) admits the stationary solution

$$\psi(z,t) = \theta(z) \exp\left[-\frac{i}{2m_0 \lambda (dt)^{(2/D_F)-1}} Et \right],$$

(4.2)

where E is the fractal energy of the plasma entity, $\theta(x)$ is the stationary state of fractal type of the plasma entity and m_0 is the fractal rest mass of the plasma entity. Then, $\theta(x)$ becomes a solution to the fractal nontemporal equation:

$$\partial_{zz} \theta(z) + \frac{1}{2m_0 \lambda^2 (dt)^{(4/D_F)-2}} (E - U)\theta(z) = 0.$$

(4.3)

Moreover, let us consider that the laser–target interaction may be interpreted as a spontaneous symmetry-breaking phenomenon. Then, the U potential can be explicitly given, as depicted in Figure 4.6.

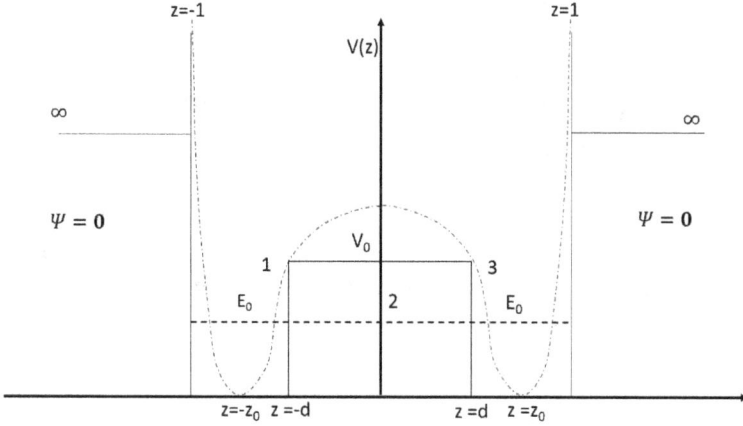

Figure 4.6. The effective potential in the case of the tunneling effect of fractal type for a physical system with spontaneous breaking symmetry (where E_0 is the fundamental energy level, V_0 is the height of the potential barrier, $z = 2d$ is the width of the potential barrier and $z = l - d$ is the width of the potential well).

In these conditions, (4.3) takes the form

$$\frac{d^2\theta_\alpha}{dz^2} + \frac{1}{2m_0\lambda^2(dt)^{\left(\frac{4}{D_F}\right)-2}}[E - V_\alpha]\theta_\alpha = 0, \quad \alpha = \overline{1,3}. \qquad (4.4)$$

For each of the three regions, the solutions of (4.4) are

$$\begin{aligned}
\theta_1(z) &= C_+e^{ikz} + C_-e^{-ikz}, \\
\theta_2(z) &= Be^{qz} + Ce^{-qz}, \\
\theta_3(z) &= D_+e^{ikz} + D_-e^{-ikz},
\end{aligned} \qquad (4.5)$$

with

$$k = \left[\frac{E}{2m_0\lambda^2(dt)^{\left(\frac{4}{D_F}\right)-2}}\right]^{\frac{1}{2}},$$

$$q = \left[\frac{V_0 - E}{2m_0\lambda^2(dt)^{\left(\frac{4}{D_F}\right)-2}}\right]^{\frac{1}{2}}. \qquad (4.6)$$

and C_+, C_-, B, C, D_+, D_- integration constants.

Due to the infinite potential in the two extreme regions, $|z| > l$, the state function of fractal type $z = +l$ implies

$$\theta_2(-l) = C_+ e^{-ikl} + C_- e^{ikl} = 0,$$
$$\theta_3(l) = D_+ e^{ikl} + D_- e^{-ikl} = 0.$$
(4.7)

Since the states density of fractal type $|\psi|^2$ is not altered by the multiplication of the state function of fractal type in the form of a constant phase factor of fractal type, the two equations for $C_\pm$ and $D_\pm$ can be immediately solved by imposing the forms

$$C_+ = \frac{A}{2i} e^{ikl}, \quad C_- = -\frac{A}{2i} e^{-ikl},$$
$$D_+ = \frac{D}{2i} e^{-ikl}, \quad D_- = -\frac{D}{2i} e^{ikl}$$
(4.8)

so that $\theta_{1,3}$ are given through simple expressions:

$$\theta_1(z) = A \sin[k(z + l)],$$
$$\theta_3(z) = D \sin[k(z - l)].$$
(4.9)

These, along with ψ_2, lead to the concrete form of "alignment conditions" in $z = \pm d$:

$$\theta_1(-d) = \theta_2(-d), \quad \theta_2(d) = \theta_3(d)$$
$$\frac{d\theta_1}{dz}(-d) = \frac{d\theta_2}{dz}(-d), \quad \frac{d\theta_2}{dz}(d) = \frac{d\theta_3}{dz}(d),$$
(4.10)

namely

$$e^{-qd} B + e^{qd} C = A \sin[k(l - d)],$$
$$q e^{-qd} B - q e^{qd} C = kA \cos[k(l - d)] \text{ in } z = -d,$$
$$e^{qd} B + e^{-qd} C = -D \sin[k(l - d)],$$
$$q e^{qd} B - q e^{-qd} C = kD \cos[k(l - d)] \text{ in } z = d.$$
(4.11)

Due to the algebraic form of the two equation pairs, in order to establish the concrete expression of the "secular equation" (for eigenvalues E of the energy), $\Delta[E] = 0$, we avoid calculating the fourth-order determinant, $\Delta[k(E), q(E)]$, formed with the amplitude coefficients of fractal type, A, B, C, D, by employing the following:

We denote with ρ the ratio C/B, and we divide the first equation by the second one for each pair. This results in

$$\frac{e^{2qd}\rho + 1}{e^{2qd}\rho - 1} = -\frac{q}{k}tg[k(l - d)],$$
$$\frac{e^{-2qd}\rho + 1}{e^{-2qd}\rho - 1} = \frac{q}{k}tg[k(l - d)], \tag{4.12}$$

which leads to the equation for ρ:

$$\frac{e^{2qd}\rho + 1}{e^{2qd}\rho - 1} + \frac{e^{-2qd}\rho + 1}{e^{-2qd}\rho - 1} = 0. \tag{4.13}$$

We find

$$\rho^2 = 1,$$

which implies

$$\rho_- = -1, \quad \rho_+ = 1. \tag{4.14}$$

For $\rho_+ = 1$, the amplitude function of fractal type, $\theta_2(z) \cong \cosh(qz)$, is symmetric just as the states of fractal type with regard to the (spatial) reflectivity against the origin. Then, the permitted values equation of the fractal energy of these states, E_S, has the concrete form

$$\tan[k_S(l - d)] = -\frac{\coth(q_S d)}{q_S}k_S, \tag{4.15}$$

where

$$k_S = \left[\frac{E_S}{2m_0\lambda^2(dt)^{\left(\frac{4}{D_F}\right)-2}}\right]^{\frac{1}{2}},$$
$$q_S = \left[\frac{V_0 - E_S}{2m_0\lambda^2(dt)^{\left(\frac{4}{D_F}\right)-2}}\right]^{\frac{1}{2}}. \tag{4.16}$$

For $\rho_- = -1$, the amplitude function of fractal type $\theta_2(z) \cong \sinh(qz)$ so that the states of fractal type will be antisymmetric, and the permitted values equation, E_A, becomes

$$\tan[k_A(l - d)] = -\frac{\tanh(q_A d)}{q_A} k_A,\qquad(4.17)$$

where

$$k_A = \left[\frac{E_A}{2m_0 \lambda^2 (dt)^{\left(\frac{4}{D_F}\right)-2}} \right]^{\frac{1}{2}},$$

$$q_A = \left[\frac{V_0 - E_A}{2m_0 \lambda^2 (dt)^{\left(\frac{4}{D_F}\right)-2}} \right]^{\frac{1}{2}}.\qquad(4.18)$$

Now, some consequences are notable: the presence of the barrier (of finite height V_0) between $-d$ and d leads to the splitting of the fundamental level E_0 into two sublevels, E_S and E_A, accounting for the two states of fractal type, symmetric and antisymmetric, respectively, in which the system can be found. In the following, the above results will be calibrated to the LPP dynamics. Indeed, these results, at a global scale resolution, can be seen as a structuring of the laser ablation plasma through its separation into two dynamical modes. We distinguish a Coulomb mode, corresponding to the fast structure, and a thermal mode, corresponding to the slow structure. The identification of the plasma structures at a certain scale resolution can be performed by admitting that the quantity $\Delta E = |E_A - E_S|$ is small compared to the plasma potential E_0, i.e. $\Delta E \ll E_0$, which implies the fact that $q_{A,S}$ is very close to

$$q_0 = \left[\frac{V_0 - E_0}{2m_0 \lambda^2 (dt)^{\left(\frac{4}{D_F}\right)-2}} \right]^{\frac{1}{2}}\qquad(4.19)$$

and also considering that

$$\frac{\coth(q_0 d)}{\tanh(q_0 d)} = \left(\frac{e^{q_0 d} + 1}{e^{q_0 d} - 1} \right)^2 > 1,\qquad(4.20)$$

with $d > 0$. This results in the fast structure being induced by the antisymmetric energy state E_A while the slower structure by the symmetrical one, E_S. Let us now calibrate the theoretical model using the empirical data presented in the previous sections. Therefore:

(i) by admitting the functionality of the scale resolution superposition principle, it results in that at global scale resolution (overall plasma plume), the following fractal equations are valid:

$$E_A = 2m_0\lambda(dt)^{\left(\frac{2}{D_F}\right)-1}f_A,$$
$$E_S = 2m_0\lambda(dt)^{\left(\frac{2}{D_F}\right)-1}f_S, \tag{4.21}$$

where f_A is the characteristic frequency of the Coulomb modes and f_S is the characteristic frequency of the thermal modes. Since $E_A > E_S$, it results in that the frequency of the Coulomb mode will always be higher than that of the thermal mode, as it was shown experimentally in Figure 4.2.

(ii) Equations (4.15) and (4.17) with the restrictions

$$k_A(l-d) \ll 1, \quad q_Ad \approx q_0d \ll 1,$$
$$k_S(l-d) \ll 1, \quad q_Sd \approx q_0d \ll 1, \tag{4.22}$$
$$l \gg d$$

take the approximate form:

$$E_A \approx E_S \approx (V_0 - E_0) + \frac{6m_0\lambda^2(dt)^{\left(\frac{4}{D_F}\right)-2}}{l^2}, \tag{4.23}$$

or even a simpler form by multiplying with $1/2m_0\lambda(dt)^{\left(\frac{2}{D_F}\right)-1}$:

$$f_A \approx f_S = f \approx f_0 + \frac{3\lambda(dt)^{\left(\frac{2}{D_F}\right)-1}}{l^2}, \tag{4.24}$$

where the following notations were used:

$$f = \frac{E_A}{2m_0\lambda(dt)^{\left(\frac{2}{D_F}\right)-1}} = \frac{E_S}{2m_0\lambda(dt)^{\left(\frac{2}{D_F}\right)-1}},$$
$$f_0 = \frac{V_0 - E_0}{2m_0\lambda(dt)^{\left(\frac{2}{D_F}\right)-1}}, \tag{4.25}$$

It results in that the value of f decreases with an increase in l and increases with an increase in $(V_0 - E_0)$. In this context, if we can identify l with the target-probe distance and $(V_0 - E_0)$ by knowing the laser beam energy, the theoretical model can describe the experimental behavior of the laser produced plasmas presented in Figures 4.2.

(iii) If we now consider the plasma entities (charges, neutrals) moving on Peano-type curves (for $D_F \to 2$), f_0 takes the form $f_0 = \frac{V_0 - E_0}{2m_0\lambda}$. Accepting now that λ takes the following form: $\lambda = \bar{\lambda}v_0 = \sigma n^{-1}v_0$, where σ is the collision cross-section and n is the particle density, it now results in that the frequency increases with an increase in the laser fluence (through n, see Figure 4.2).

(iv) Considering that the energies from (4.21) are of kinetic nature, $E_A = \frac{m_0 v_A^2}{2}$ and $E_S = \frac{m_0 v_S^2}{2}$, coupled with the fact that the separation of the plasma plume into the two dynamical modes can be expanded at local scale resolutions, $\lambda_A(dt)^{\left(\frac{2}{D_F}\right)-1}$ and $\lambda_S(dt)^{\left(\frac{2}{D_F}\right)-1}$, respectively, from (4.21) it results in

$$f_A = \frac{v_A^2}{4\lambda_A(dt)^{\left(\frac{2}{D_F}\right)-1}}, \quad f_S = \frac{v_S^2}{4\lambda_S(dt)^{\left(\frac{2}{D_F}\right)-1}}, \qquad (4.26)$$

where v_A is identified as the velocity of the fast structure and v_S is the velocity of the slower one. Particularly, for the ablation plasma dynamics associated with the scale transitions, such as correlative–noncorrelative processes, the fractal dimension of the movement curves has the value $D_F = 2$. Then, (4.26) becomes

$$f_A = \frac{v_A^2}{4\lambda_A}, \quad f_S = \frac{v_S^2}{4\lambda_S}. \qquad (4.27)$$

Moreover, for

$$\lambda_A = \frac{kT_e}{m\bar{v}_A}, \quad \lambda_S = \frac{kT_e}{m\bar{v}_S}, \qquad (4.28)$$

where k is the Boltzmann constant, T_e is the electron temperature and m is the atomic mass of the ions, $\bar{v}_A$, is the ion–electron collision frequency at a Coulomb scale resolution and $\bar{v}_S$ is the collision frequency at a thermal scale resolution, (4.27) with (4.28) become

$$f_A = \frac{m\bar{v}_A v_A^2}{kT_e}, \quad f_S = \frac{m\bar{v}_S v_S^2}{kT_e}. \tag{4.29}$$

We now perform a numerical evaluation of (4.29) using experimental data as input parameters. The experimental data were presented in the previous section, and the simulated results are summarized in Table 4.1. We notice a good correlation between the experimental data and the simulated ones. Small discrepancies can be seen for the second structure at larger distances, where the experimental data are nosier, and the values can be affected by errors. Also, this representation showcases once again the versatility of the fractal theoretical approach, as once the model is calibrated to a specific measurement, it can offer results in very good agreement with the empirical data.

4.3.3. *Mutual conditionings of the plasma substructures trough joint invariant functions*

The fact that the presence of two plasma structures can be explained through the self-structuring of the plasma plume in two dynamical modes shows the mutual conditionings between these two substructures. In order to characterize these conditionings, we must admit, according to the empirical proof presented in Figure 4.1, that the two plasma substructures are characterized by two differential equations of damped oscillator type for the charged particle densities. By using the group properties of these differential equations (invariances of $SL(2\mathbb{R})$ type, algebras of Lie type, etc., (Eason, 2006) and the Stoka theorem (Stoka, 1967), we build the joint invariant functions on the basis of two isomorphous $SL(2\mathbb{R})$ groups. These functions will describe the mutual conditionings between the two plasma substructures. Following this, let us consider the Lie algebra base associated with the $SL(2\mathbb{R})$ group given by the infinitesimal generators

$$A_1 = \frac{\partial}{\partial h} + \frac{\partial}{\partial \bar{h}}, \quad A_2 = h\frac{\partial}{\partial h} + \bar{h}\frac{\partial}{\partial \bar{h}}, \quad A_3 = h^2\frac{\partial}{\partial h} + \bar{h}^2\frac{\partial}{\partial \bar{h}} \tag{4.30}$$

Table 4.1. Comparison between the experimental and theoretical values of the ionic frequency for the first (a) and second (b) plasma structures.

(a)

Fluence (J/cm²)	First structure experimental data (MHz)				First structure simulated data (MHz)			
	1 cm	2 cm	2.5 cm	3 cm	1 cm	2 cm	2.5 cm	3 cm
28	17.5 ± 0.2	15 ± 0.8	13 ± 0.5	7.6 ± 0.6	18.5 ± 0.3	16.1 ± 0.1	13.5 ± 0.	9.4 ± 0.2
57	19.5 ± 0.3	16.8 ± 0.6	13.5 ± 0.7	8.5 ± 0.7	21.7 ± 0.2	18.2 ± 0.4	14.2 ± 0.6	10.6 ± 0.35
85	21 ± 0.7	19 ± 0.4	18 ± 0.5	16.3 ± 0.1	22.4 ± 0.05	20.2 ± 0.6	17.7 ± 0.5	16.3 ± 0.2
115	22 ± 0.1	19.26 ± 0.2	18.5 ± 0.3	17.3 ± 0.2	22.8 ± 0.5	20.9 ± 0.3	18.4 ± 0.2	17.4 ± 0.1

(b)

Fluence (J/cm²)	Second structure experimental data (MHz)				Second structure simulated data (MHz)			
	1 cm	2 cm	2.5 cm	3 cm	1 cm	2 cm	2.5 cm	3 cm
28	7.8 ± 0.1	6.5 ± 0.3	2 ± 0.4	1.2 ± 0.6	7.4 ± 0.6	6.44 ± 0.1	2.4 ± 0.3	2.2 ± 0.1
57	8.4 ± 0.2	7.2 ± 0.1	5.6 ± 0.1	4.58 ± 0.2	8.68 ± 0.6	7.28 ± 0.04	5.68 ± 0.05	4.35 ± 0.05
85	9.5 ± 0.4	9.3 ± 0.5	9 ± 0.3	8.2 ± 0.2	8.96 ± 0.6	9.08 ± 0.1	8.78 ± 0.04	8.14 ± 0.04
115	10 ± 0.5	9.8 ± 0.5	9.3 ± 0.4	8.6 ± 0.3	9.12 ± 0.6	9.36 ± 0.06	8.99 ± 0.05	8.4 ± 0.2

and its structure

$$[A_1, A_2] = A_1, \quad [A_2, A_3] = A_3, \quad [A_3, A_1] = -2A_2, \qquad (4.31)$$

where h is a complex amplitude and $\bar{h}$ is its complex conjugate. We admit that this group would characterize the Coulomb structure behavior.

Let us also consider the Lie algebra base associated with the $SL(2\mathbb{R})$ group given by the infinitesimal generators

$$B_1 = \frac{\partial}{\partial y} + \frac{\partial}{\partial \bar{y}}, \quad B_2 = y\frac{\partial}{\partial y} + \bar{y}\frac{\partial}{\partial \bar{y}}, \quad B_3 = y^2\frac{\partial}{\partial y} + \bar{y}^2\frac{\partial}{\partial \bar{y}} \qquad (4.32)$$

and its structure

$$[B_1, B_2] = B_1, \quad [B_2, B_3] = B_3, \quad [B_3, B_1] = -2B_2, \qquad (4.33)$$

where y is a complex amplitude and $\bar{y}$ is its complex conjugate. We admit now that this group would characterize the thermal structure behavior.

With the above-mentioned groups being isomorphous, the joint invariant functions F at the actions of these groups will be obtained as solutions to Stoka's (Stoka, 1967) equations:

$$A_l F + B_l F = 0, \quad l = 1, 2, 3, \qquad (4.34)$$

or in a more explicit way,

$$\frac{\partial F}{\partial h} + \frac{\partial F}{\partial \bar{h}} + \frac{\partial F}{\partial y} + \frac{\partial F}{\partial \bar{y}} = 0,$$

$$h\frac{\partial F}{\partial h} + \bar{h}\frac{\partial F}{\partial \bar{h}} + y\frac{\partial F}{\partial y} + \bar{y}\frac{\partial F}{\partial \bar{y}} = 0, \qquad (4.35)$$

$$h^2\frac{\partial F}{\partial h} + \bar{h}^2\frac{\partial F}{\partial \bar{h}} + y^2\frac{\partial F}{\partial z} + \bar{y}^2\frac{\partial F}{\partial \bar{y}} = 0.$$

The rank's system is three; therefore, it exists as only one independent integral. This is a cross-ratio generated by means of the relation

$$\frac{h - y}{h - \bar{y}} : \frac{\bar{h} - y}{\bar{h} - \bar{y}} = \sigma, \qquad (4.36)$$

where σ is real and positive. Now, any joint invariant function is here a regular function of this cross-section. In particular, for: $\sigma = \tanh\tau$, with τ arbitrary, through (4.36), z is connected to h by means of the relation

$$y = u + vh_0, \tag{4.37}$$

with

$$h = u + iv,$$
$$h_0 = -i\frac{\cosh\tau - e^{-i\alpha}\sinh\tau}{\cosh\tau + e^{-i\alpha}\sinh\tau}, \tag{4.38}$$
$$\Delta\tau = 0,$$

where Δ is the Laplace operator and α is real. Equations (4.37) and (4.38) establish a relation between the current density amplitude of the two-plasma structure. Moreover, through (4.38), we showcase a self-modulation in the amplitude of the ionic current density through a Stoler-type transformation. We showcase Stoler-type transformations with an important role in the theory of coherent states, meaning in the charge creation or annihilation processes. The possibility of using Stoler-type transformation for the analysis of laser-produced plasma dynamics has more profound implications. The existence of coherent states implies that the individual ejected particles are coherent and that there is a connecting tissue among the ablated particle cloud. This aspect of the laser-produced plasma was empirically showcased by our group when we showed that at all laser ablation scales, there are mathematical functions connecting the series of materials, as the fractal model already encompasses. Since the particularity of a fractal model is the use of zoom in/zoom out processes, we can still find a coherence between the plasma substructures (Coulomb or thermal). This is seen experimentally, as although the individual values differ for each structure, the overall trends are respected regardless of the nature of the substructure. Another important role is played by the creation–annihilation processes. These processes correspond to the ionization–neutralization processes occurring during the expansion of the plasma. The ionization processes are directly connected to the collisions during expansion, which are accounted for in the framework of our fractal model through the fractalization degree. Our mathematical approach is, therefore, a complex one, accounting for the main processes occurring during the expansion of the laser-produced plasma, and through an adequate calibration of the model, we can offer a quantitative analysis of the charges modulated dynamics.

4.3.4. Complex dynamics through Schrödinger "regimes" of multifractal type

"Hidden symmetry" in transient plasma dynamics

In the one-dimensional stationary case, the Schrödinger multifractal-type equation (see Chapter 3) takes the form

$$\frac{d^2\Psi}{dx^2} + k_0^2\Psi = 0, \tag{4.39}$$

with

$$k_0^2 = \frac{E}{2m_0\lambda^2(dt)^{\left[\frac{4}{f(\alpha)}\right]-2}}. \tag{4.40}$$

In (4.40), E is the multifractal energy of the transient plasmas structural unit and m_0 is the rest mass of the transient plasmas structural unit.

The solution to (4.39) can be written in the form

$$\Psi(x) = he^{i(k_0x+\theta)} + \acute{h}e^{-i(k_0x+\theta)}, \tag{4.41}$$

where h is the complex amplitude, $\acute{h}$ is the complex conjugate of h and θ is a phase. Thus, $h, \acute{h}$ and θ label each structural unit from transient plasmas that have as a "fundamental property" the same k_0.

Equation (4.39) has a "hidden" symmetry by means of a homographic group of multifractal type. Indeed, the ratio ε of two independent linear solutions to (4.39) is a solution to Schwartz's differential equation of multifractal type (Mihaileanu, 1976):

$$\{\varepsilon, x\} = \frac{d}{dx}\left(\frac{\ddot{\varepsilon}}{\dot{\varepsilon}}\right) - \frac{1}{2}\left(\frac{\ddot{\varepsilon}}{\dot{\varepsilon}}\right)^2 = 2k_0^2, \tag{4.42}$$

$$\dot{\varepsilon} = \frac{d\varepsilon}{dx}, \quad \ddot{\varepsilon} = \frac{d^2\varepsilon}{dx^2}. \tag{4.43}$$

The left-hand side of (4.42) is invariant with respect to the homographic transformations of multifractal type:

$$\varepsilon \leftrightarrow \varepsilon' = \frac{a\varepsilon + b}{c\varepsilon + d}, \tag{4.44}$$

with a, b, c and d being real parameters (of multifractal type). The relation (4.44) corresponding to all possible values of these parameters defines the group $SL(2\mathbb{R})$ of multifractal type.

Thus, all the transient plasma's structural units having the same k_0 are in biunivocal correspondence with the transformations of the group $SL(2\mathbb{R})$ of the multifractal type. This allows the construction of a "personal" parameter of multifractal type, ε, for each transient plasma's structural unit, separately. Indeed, as a "guide", it is chosen as the general form of the solution to (4.42), which is written as

$$\varepsilon' = l + m \quad tan(k_0 x + \theta). \tag{4.45}$$

Thus, using l, m and θ, it is possible to characterize any transient plasma's structural unit. In such a conjecture, identifying the phase from (4.45) with the one from (4.41), the "personal" parameter of multifractal type becomes

$$\varepsilon' = \frac{h + \bar{h}\varepsilon}{1 + h}, \quad h = l + \Im m, \quad \bar{h} = l - \Im, \quad m\varepsilon \equiv e^{2i(k_0 x + \theta)}. \tag{4.46}$$

The fact that (4.45) is also a solution to (4.42) implies, by explaining (4.44), the group of $SL(2\mathbb{R})$ of multifractal type (Merches and Agop, 2016; Agop and Paun, 2017):

$$h' = \frac{ah + b}{ch + d}, \quad \bar{h} = \frac{a\bar{h} + b}{c\bar{h} + d}, \quad k' = \frac{c\bar{h} + d}{ch + d}k. \tag{4.47}$$

Therefore, the group (4.47) works as "synchronization modes" among the various structural units of any transient plasma process, to which the amplitudes and phases of each of them obviously participate, in the sense that they are also connected. More precisely, through the group (4.47), the phase of k is only moved by a quantity depending on the amplitude of the plasma's structural units at the transition among various structural units of any transient plasma. Furthermore, the amplitude of the structural unit of any transient plasma is also affected from a homographic perspective. The usual "synchronization" manifested through the delay of the amplitudes and phases of the structural units of any transient plasma must represent here a particular case.

The structure of group (4.38) is typical of $SL(2\mathbb{R})$, which will be taken in the standard form:

$$[A_1, A_2] = A_1, \quad [A_2, A_3] = A_3, \quad [A_3, A_1] = -2A_2, \quad (4.48)$$

where $A_k, k = 1, 2, 3$ are the infinitesimal generators of the group. Because the group is simple transitive, these generators can be easily found as the components of the Cartan conframe of multifractal type from the relation

$$d(f) = \sum \frac{\partial f}{\partial x^k} dx^k = \left\{ \omega^1 \left[h^2 \frac{\partial}{\partial h} + \bar{h}^2 \frac{\partial}{\partial \bar{h}} + (h - \bar{h})k \frac{\partial}{\partial k} \right] \right. $$
$$\left. + 2\omega^2 \left(h \frac{\partial}{\partial h} + \bar{h} \frac{\partial}{\partial \bar{h}} \right) + \omega^3 \left(\frac{\partial}{\partial h} + \frac{\partial}{\partial \bar{h}} \right) \right\} (f), \quad (4.49)$$

where ω^k are the components of the Cartan coframe of multifractal type to be found from the system

$$dh = \omega^1 h^2 + 2\omega^2 h + \omega^3, \quad d\bar{h} = \omega^1 \bar{h}^2 + 2\omega^2 \bar{h} + \omega^3,$$
$$dk = \omega^1 k(h - \bar{h}). \quad (4.50)$$

Thus, both the infinitesimal generators and the coframe of the multifractal types are obtained by identifying the right-hand side of (4.49) through the standard dot product of $SL(2\mathbb{R})$ algebra of multifractal type,

$$\omega^1 A_3 + \omega^3 A_1 - 2\omega^2 A_2, \quad (4.51)$$

so that

$$A_1 = \frac{\partial}{\partial h} + \frac{\partial}{\partial \bar{h}}, \quad A_2 = h \frac{\partial}{\partial h} + \bar{h} \frac{\partial}{\partial \bar{h}},$$
$$A_3 = h^2 \frac{\partial}{\partial h} + \bar{h}^2 \frac{\partial}{\partial \bar{h}} + (h - \bar{h}) k \frac{\partial}{\partial k} \quad (4.52)$$

and

$$\omega^1 = \frac{dk}{(h - \bar{h})k}, \quad 2\omega^2 = \frac{dh - d\bar{h}}{h - \bar{h}} - \frac{h + \bar{h}}{h - \bar{h}}\frac{dk}{k},$$
$$\omega^3 = \frac{hdh - \bar{h}d\bar{h}}{h - \bar{h}} + \frac{h\bar{h}dk}{(h - \bar{h})k}. \tag{4.53}$$

It is worth mentioning that this does not work with the previous differential forms but with the absolute invariant differentials

$$\omega^1 = \frac{dh}{(h - \bar{h})\,k}, \quad \omega^2 = -i\left(\frac{dk}{k} - \frac{dh + d\bar{h}}{h - \bar{h}}\right), \quad \omega^3 = \frac{kd\bar{h}}{h - \bar{h}}. \tag{4.54}$$

The advantage of this representation is that it makes obvious the connection with the Poincaré representation of the Lobachevsky plane. Indeed, the metric here is

$$\frac{ds^2}{g} = (\omega^2)^2 - 4\omega^1\omega^2 = \left(\frac{dk}{k} - \frac{dh + d\bar{h}}{h - \bar{h}}\right)^2 + 4\frac{dhd\bar{h}}{(h - \bar{h})^2}, \tag{4.55}$$

where g is a constant.

These metrics reduce to those of Poincaré in the case when $\omega^2 = 0$, which defines the variable θ as the "angle of parallelism" (in the Levi–Civita sense) of the hyperbolic plane of multifractal type (the connection of the multifractal type).

"Synchronization modes" through Riccati-type gauge in transient plasma dynamics

Returning to the homographic transformation of multifractal type (4.44), according to the previously presented implications of this transformation, each structural unit of any transient plasma can be located either by using homogenous coordinates (a, b, c, d) or three non-homogenous coordinates when a parallelism of direction in the Levi–Civita sense becomes functional on the manifold induced by the $SL(2\mathbb{R})$ group of multifractal type. Now, the simultaneity condition of the free structural units of any transient plasma can be differently characterized from a Riccati equation of multifractal type in pure differentials of multifractal type (this is named Riccati gauge of multifractal type):

$$d\frac{a\varepsilon + b}{c\varepsilon + d} = 0, \tag{4.56}$$

which implies

$$d\varepsilon = w^1\varepsilon^2 + w^2\varepsilon + w^3, \tag{4.57}$$

where w^1, w^2 and w^3 are the components of the Cartan coframe of multifractal type obtained through (4.53). Therefore, for the description of any transient plasma dynamics as a succession of states of an ensemble of simultaneous structural units, as it were, it suffices to have three differentiable 1- forms, representing a coframe of $SL(2\mathbb{R})$ algebra of multifractal type. Consequently, a state of a transient plasma in a given dynamics can be organized as a metric plane space, i.e. a Riemannian three-dimensional space of multifractal type. Accordingly, the geodesics of a Riemannian space of multifractal type are given by certain conservations of the equations of multifractal type:

$$w^1 = a^1 d\tau, \quad w^2 = a^2 d\tau, \quad w^3 = a^3 d\tau, \tag{4.58}$$

where a^1, a^2 and a^3 are constants and τ is the affine parameter of the geodesics, so that along these geodesics of differential equation, (4.57) is an ordinary differential of Riccati type:

$$\frac{d\varepsilon}{d\tau} = a^1\varepsilon^2 + 2a^2\varepsilon + a^3. \tag{4.59}$$

Let us consider the following form of the previous equation:

$$A\frac{d\varepsilon}{d\tau} - \varepsilon^2 + 2B\varepsilon + AC = 0, \tag{4.60}$$

where

$$\frac{1}{a^1} = A, \quad -2\frac{a^2}{a^1} = B, \quad -\frac{a^3}{a^1} = AC. \tag{4.61}$$

Since the roots of the polynomial

$$P(\varepsilon) = \varepsilon^2 - 2B\varepsilon - AC \tag{4.62}$$

can be written in the form

$$\varepsilon_1 = B + iA\Omega, \quad \varepsilon_2 = B - iA\Omega, \quad \Omega^2 = \frac{C}{A} - \left(\frac{B}{A}\right)^2, \tag{4.63}$$

the change of variable

$$z = \frac{\varepsilon - \varepsilon_1}{\varepsilon - \varepsilon_2} \tag{4.64}$$

transforms into

$$\dot{z} = 2i\Omega z \tag{4.65}$$

Of the solution

$$z(\tau) = z(0)e^{2i\Omega\tau}. \tag{4.66}$$

Therefore, if the initial condition $z(0)$ is conveniently expressed, then it is possible to construct the general solution to (4.59) by writing the transformation (4.64) in the form

$$\varepsilon = \frac{\varepsilon_1 + re^{2i\Omega(\tau-\tau_0)}\varepsilon_2}{1 + re^{2i\Omega(\tau-\tau_0)}}, \tag{4.67}$$

where r and τ_0 are two integration constants. Using (4.63), it is possible to write this solution in real terms:

$$z = B + A\Omega \left\{ \frac{2r\sin\left[2\Omega(\tau-\tau_0)\right]}{1 + r^2 + 2r\cos\left[2\Omega(\tau-\tau_0)\right]} \right.$$

$$\left. + i\frac{1 - r^2}{1 + r^2 + 2r\cos\left[2\Omega(\tau-\tau_0)\right]} \right\}. \tag{4.68}$$

Therefore, "synchronization modes" in the phase and amplitude of the plasma structural units imply group invariances of $SL(2\mathbb{R})$ type. Then, period doubling, quasi-periodicity, intermittence, etc., emerge as natural behaviors in the transient plasma dynamics (see Figures 4.7(a)–(p) for $r = 0.5$ and real $[(z - B)/A] \equiv$ amplitude at various scale resolutions, given by means of the maximum value of Ω).

As it can be observed in Figures 4.7(a)–(p), the natural transition of a transient plasma is to evolve from a normal period doubling state toward damped and a strong modulated dynamic. The transient plasma never reaches a chaotic state, but it always evolves toward that state. The transient plasma jumps directly to a doubling-period state, and thus, it again follows the same scenario presented here.

The evolution of the transient plasma is further studied with an increase in the control parameter. To this end, for a relatively small

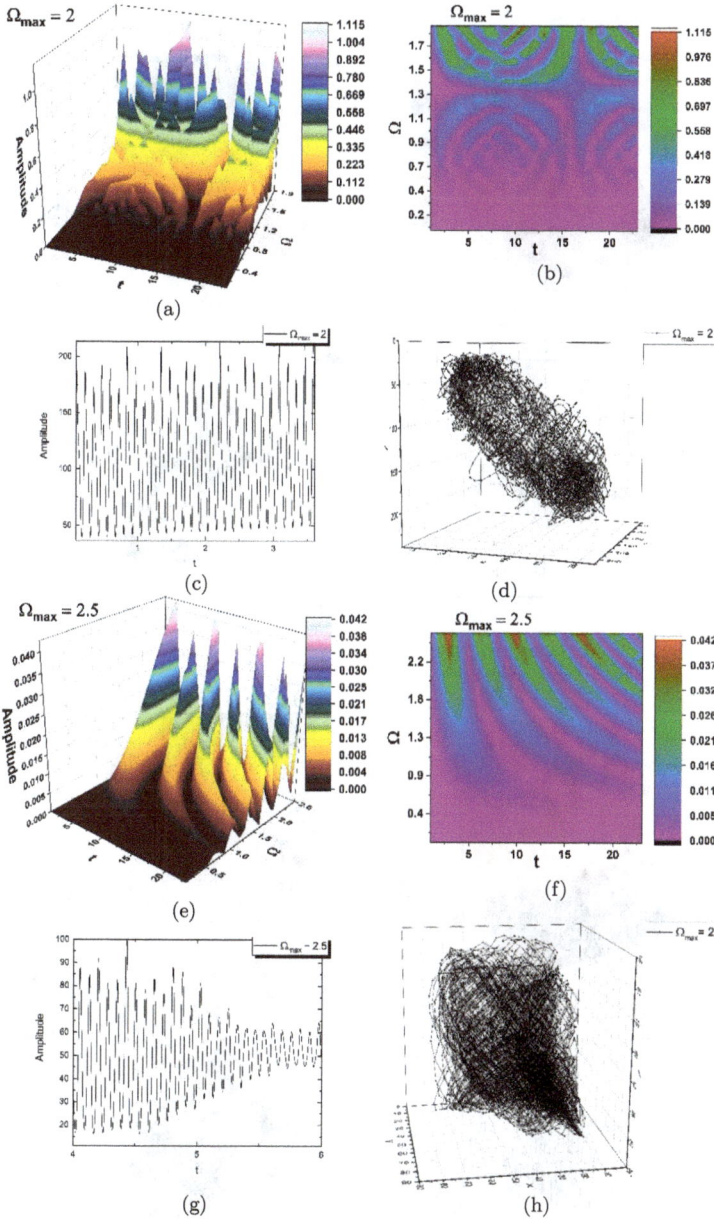

Figure 4.7. (a)–(p) Various "synchronization modes" of the transient plasma's structural units (3D, contour plot, time series and reconstructed attractors for various modes of the scale resolution given by Ω_{max}); period doubling (a, b, c, d); damping regimes (e, f, g, h); quasi-periodicity (i, j, k, l); intermittences (m, n, o, p).

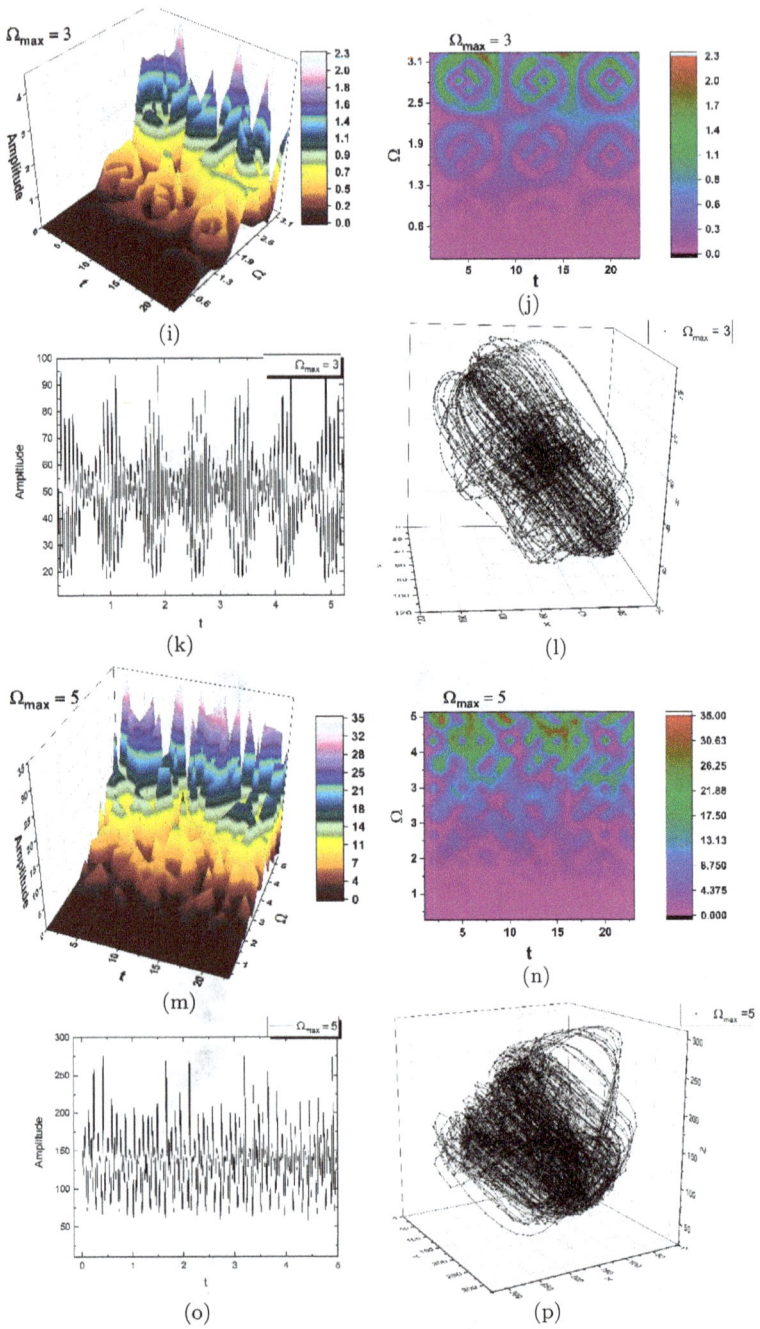

(i)

(j)

(k)

(l)

(m)

(n)

(o)

(p)

Figure 4.7. (*Continued*)

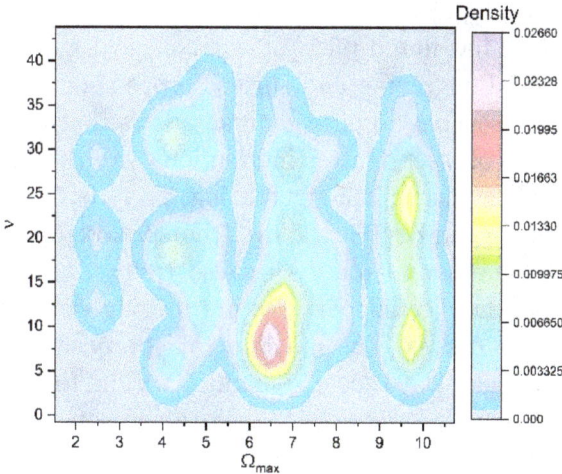

Figure 4.8. Frequency of the charged plasma particles as a function of a scale resolution chosen by Ω_{max}: Bifurcation map.

range of values, the response of the transient plasma is investigated. It is observed (see Figures 4.7(a)–(p)) that the plasma particle dynamics starts from a double-period state, transitions to a damped state and evolves through a quasi-chaotic state, which is never achieved. The transition is evidenced by the presence of supplementary frequencies. While the frequency response of the plasma charged particle is somehow periodic, the amplitude increases quasi-linearly as the values of the control parameters increase. The bifurcation map is presented in Figure 4.8, where again it is observed that the plasma charged particle dynamics starts from a steady state (double-period state) and evolves toward a chaotic one ($\Omega_{max} = 2, 2.5, 3, \ldots$), but it never reaches that state.

4.3.5. *Implementation of a multifractal theoretical model for laser-induced plasmas*

The use of multifractal models to describe the complex dynamics of low-temperature plasmas has been the trademark of our group in the past few years, where we have looked into either the nonlinear behavior of self-organized structures or the behavior of laser-produced plasmas. The latter problem has become more interesting, as it has found relevant technological applications, such as PLD, material engraving and surface processing. The aim of the multifractal approach is to offer generalist laws that could unify the different aspects of laser–matter interactions. We previously tackled processes such as plasma self-structuring, particle distribution in a single element and in complex alloys, the spatiotemporal evolution of certain plasma parameters. All these results were achieved by admitting that the ejected particles are moving on continuous but nondifferentiable curves. The multifractal approach allows an easy transition from global to local dynamics within the framework of the same mathematical model. This is the reason why this rather difficult theory offers the best chance of finding general laws that could even showcase the laser beam — target — plasma relationship and thin-film properties.

When analyzing the dynamics of laser-produced plasmas as a complex system in a multifractal theoretical construction, we can focus on all facets of the LPP, including its optical emission, global dynamics, molecular and cluster ejection and ion or electron kinetics. Out of all the possible directions, the one that more truly embeds the multifractal behavior is the analysis of the electron dynamics. Due to their lower mass, they are ejected from the target via several ablation mechanisms and suffer multiple collisions, which lead to ionization, neutralization and excitation processes. Basically, the electrons act as the multifractal matrix of the LPP dynamics. In order to implement the theoretical consideration presented in the previous section in describing the LPP dynamics, we choose to investigate the transient plasmas generated by UV–nanosecond-laser ablation of nickel (Ni) and magnesium (Mg) samples. The experiments were performed in a stainless-steel vacuum chamber pumped down to 2×10^{-5} Torr residual pressure. The radiation from a Nd-YAG nanosecond laser (355 nm, third harmonic, pulse width = 5 ns, variable fluences) was focused by a f = 30 cm lens onto Mg and Ni targets (the spot diameter at the impact point was approximately 0.3 mm) placed in the vacuumed chamber. The metallic targets rotated during the experiments and were grounded from the vacuum chamber. Before each measurement, a surface cleaning procedure was implemented in order to remove the oxide layer present on the surface of the target. The total ionic and electronic currents extracted from the plasma plume were measured using a tungsten heated cylindrical LP with 0.25 mm diameter and 3 mm length. The probe was heated in order to avoid deposition of the target material on the probe, thus keeping constant the collecting area of the probe during measurements. The LP was also biased with a voltage of +5 V using a stabilized DC power source.

In Figure 4.9, we represent the electronic temporal traces collected on the main expansion axis at 3 cm from the target. We observe three

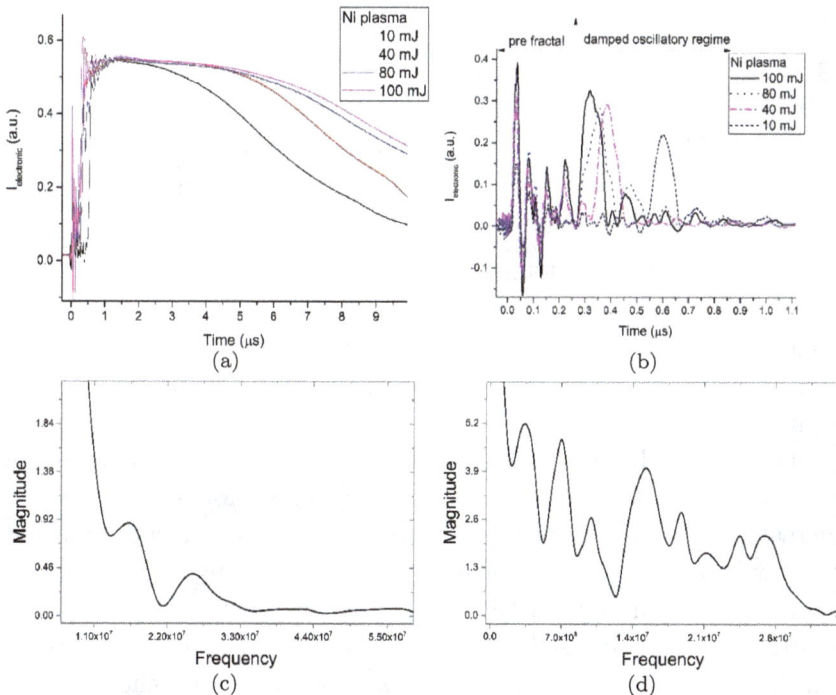

Figure 4.9. Representation of the saturation electronic current (a), dynamical regimes (b), the FFT of the pre-fractal area (c) and the damped regime (d) for the Ni plasma, and the saturation electronic current (e), modulated regime (f), the FFT of the pre-fractal area (g) and the damped regime (h) for the Mn plasma.

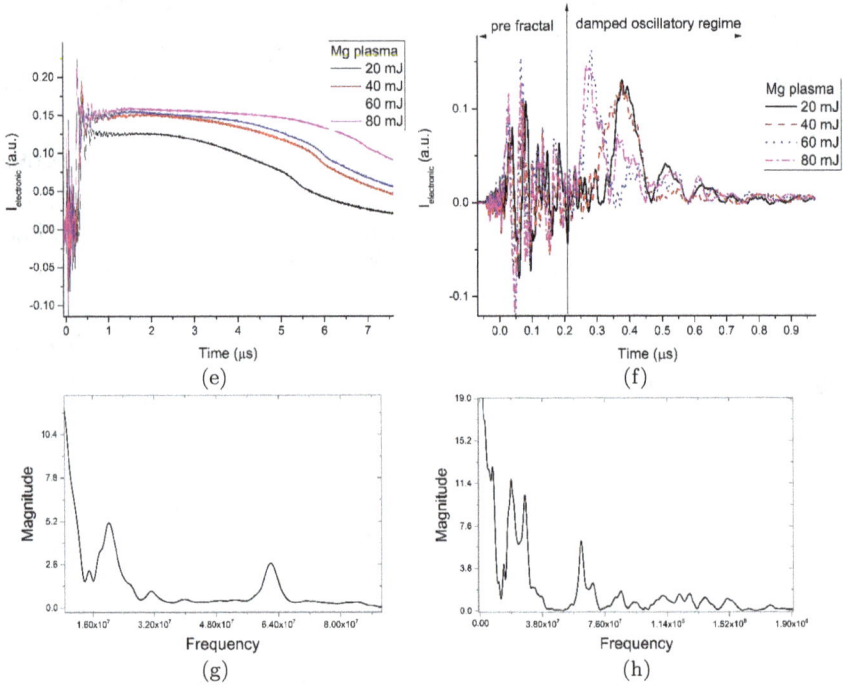

Figure 4.9. (*Continued*)

distinct areas: The first area is a mixed ionic and electronic complex multi-peak regime, followed by a convoluted electronic damped regime and a well-known shifted Maxwell–Boltzmann distribution of the current. In the incipient region, we can see that it does not depend on the laser energy, as the only effect found was a slight increase of 10% in the overall collected electronic current. However, we have found a deeper meaning for this "strange" region. This regime is different for the Mg and Ni plasmas; therefore, it is specific to the type of target used. In the paradigm of our model, this region, which is constant regardless of the laser parameter but changes with the nature of the target, is indeed a pre-fractal. As we can see from the FTT, the electron dynamics can be described by several frequencies, but they are not chaotic. This region acts as a projection in the differentiable, measurable dynamics of a multidimensional multifractal generated for each individual target. This projection represents a wide range of possibilities for the ejected electrons to follow as the dynamics are "chosen". This pre-fractal state of the plasma is different for the Mg and Ni plasmas since it reflects more of the properties of the target, as the plasma has not yet transformed into a nonequilibrium soup of high-energetic electron ions and localized optical emission. With further development of the electronic cloud, we see a damped regime, which is described by two frequencies. These frequencies depend on the properties of the laser, measurement angle and the distance as the plasma state of the ablated cloud is clearly formed. The recorded frequencies are part of the multitude of initial possible states, from which two are selected based on the geometry and the experimental conditions. This is the first proof of the fractal theory, as the selection of the dynamics can be seen empirically.

The experimental data presented here showcase an interesting scenario, in which metallic ablation plasmas transition from a quasi-chaotic state into a simple double-period damped dynamic. The mathematical model predicts such a behavior, as the sequence presented in Figure 4.8 is a cyclical one and the system always reverts to the double period. It is worth noting that the theoretical model predicts a transition from a double-period state to a damped state, with both dynamics confirmed experimentally. As for the multi-structuring, it is not clearly seen here because it occurs at a considerably later evolution time of the plasma, as was seen experimentally through optical emission spectroscopy. Therefore, laser ablation plasmas are perfect media, which can behave according to the predictions made by our model, and are especially systems that don't reach a chaotic state, at least not in high-vacuum conditions, under which our experiments were performed.

References

Aragón C. an Aguilera J. A. 2008. Characterization of laser induced plasmas by optical emission spectroscopy: A review of experiments and methods, *Spectrochim. Acta Part B: At. Spec.*, 63(9), 893–916.

Bruzzese R., Spinelli N., and Velotta R. 1996. Laser produced plasmas in high fluence ablation of metallic surfaces probed by time-of-flight mass spectrometry, *Appl. Surf. Sci.*, 98, 175–180.

Bulai G., Trandafir V., Irimiciuc S. A., Ursu L., Focsa C., and Gurlui S. 2019. Influence of rare earth addition in cobalt ferrite thin films obtained by pulsed laser deposition, *Ceram. Int.*, 45(16), 20165–20171.

Bulgakov A. V. and Bulgakova N. M. 1999. Dynamics of laser-induced plume expansion into an ambient gas during film deposition, *J. Phys. D: Appl. Phys.*, 28(8), 1710–18.

Canulescu S., Papadopoulou E. L., Anglos D., Lippert Th., Schneider C. W., and Wokaun A. 2009. Mechanisms of the laser plume expansion during the ablation of LiMn2O4, *J. Appl. Phy.*, 105(6), 063107.

Chen J., Lippert T., Ojeda G.-P. A., Stender D., Schneider C. W., and Wokaun A. 2014. Langmuir probe measurements and mass spectrometry of plasma plumes generated by laser ablation of La0.4Ca0.6MnO3, *J. Appl. Phy.*, 116(7), 073303.

Craciun D., Socol G., Stefan N., Dorcioman G., Hanna M., Taylor C. R., Lambers E., and Craciun V. 2014. The effect of deposition atmosphere on the chemical composition of TiN and ZrN thin films grown by pulsed laser deposition, *Appl. Surf. Sci.*, 302, 124–28.

Craciun V. and Craciun D. 1992. Reactive pulsed laser deposition of TiN, *Appl. Surf. Sci.*, 54(1), 75–77.

Dijkkamp D., Venkatesan T., Wu X. D., Shaheen S. A., Jisrawi N., Min-Lee Y. H., McLean W. L., and Croft M. 1987. Preparation of Y-Ba-Cu oxide superconductor thin films using pulsed laser evaporation from high Tc bulk material, *Appl. Phys. Lett.*, 51(8), 619.

Doggett B. and Lunney J. G. 2011. Expansion dynamics of laser produced plasma, *J. Appl. Phy.*, 109(9).

Donnelly T., Lunney J. G., Amoruso S., Bruzzese R., Wang X., and Ni X. 2010a. Dynamics of the plumes produced by ultrafast laser ablation of metals, *J. Appl. Phys.*, 108(4).

Donnelly T., Lunney J. G., Amoruso S., Bruzzese R., Wang X., and Phipps C. 2010b. Plume dynamics in femtosecond laser ablation of metals, *AIP Conf. Procc.*, 643, 643–655.

Dorcioman G., Socol G., Craciun D., Argibay N., Lambers E., Hanna M., Taylor C. R., and Craciun V. 2014. Applied surface science wear tests of ZrC and ZrN thin films grown by pulsed laser deposition, *Appl. Surf. Sci.*, 306, 33–36.

Eliezer S. and Hora H. 1989. Surface waves in laser-produced plasma double layers, *IEE Trans. Plasma Sci.*, 17, 0–4.

Focsa C., Gurlui S., Nica P., Agop M., and Ziskind M. 2017. Plume splitting and oscillatory behavior in transient plasmas generated by high-fluence laser ablation in vacuum, *Appl. Surf. Sci.*, 424, 299–309.

Focsa C., Nemec P., Ziskind M., Ursu C., Gurlui S., and Nazabal V. 2009. Laser ablation of AsxSe100−x chalcogenide glasses: Plume investigations, *Appl. Surf. Sci.*, 255(10), 5307–5311.

Geohegan D. B. and Puretzky A. A. 1996. Laser ablation plume thermalization dynamics in background gases: Combined imaging, optical absorption and emission spectroscopy, and ion probe measurements, *Appl. Surf. Sci.*, 96–98, 131–38.

Glavin N. R., Muratore C., Jespersen M. L., Hu J., Fisher T. S., and Voevodin A. A. 2015. Temporally and spatially resolved plasma spectroscopy in pulsed laser deposition of ultra-thin boron nitride films, *J. Appl. Phys.* 117(16).

Gurlui S. M., Sanduloviciu C., Mihesan, Z. M., and Focsa C. 2009. Periodic phenomena in laser-ablation plasma plumes: A self-organization scenario, *AIP Conf. Proc.*, 821(1), 279–82.

Harilal S. S., Bindhu C. V., Tillack M. S., Najmabadi F., and Gaeris A. C. 2003. Internal structure and expansion dynamics of laser ablation plumes into ambient gases, *J. Appl. Phys.*, 93(5), 2380–2388.

Harilal S. S., Farid N., Freeman J. R., Diwakar P. K., LaHaye N. L., and Hassanein A. 2014. Background gas collisional effects on expanding Fs and Ns laser ablation plumes, *Appl. Phys. A* 117(1), 319–326.

Harilal S. S., Issac R. C., Bindhu C. V., Nampoori V. P. N., and Vallabhan C. P. G. 1996. Temporal and spatial evolution of C2 in laser induced plasma from graphite target, *J. Appl. Phys.*, 80(6), 3561.

Irimiciuc S. A., Boidin R., Bulai G., Gurlui S., Nemec P., Nazabal V., and Focsa C. 2017a. Laser ablation of (GeSe2)100−x(Sb2Se3)x chalcogenide glasses: Influence of the target composition on the plasma plume dynamics, *Appl. Surf. Sci.*, 418, 594–600.

Irimiciuc S. A., Bulai G., Agop M., and Gurlui S. 2018a. Influence of laser-produced plasma parameters on the deposition process: In situ space- and time-resolved optical emission spectroscopy and fractal modeling approach, *Appl. Phys. A: Mat. Sci. Process.*

Irimiciuc S. A., Mihaila I., and Agop M. 2014a. Experimental and theoretical aspects of a laser produced plasma, *Phys. Plasmas*, 21(9), 093509.

Irimiciuc S. A., Agop M., Nica P., Gurlui S., Mihaileanu D., Toma S., and Focsa C. 2014b. Dispersive effects in laser ablation plasmas, *Jpn. J. Appl. Phys.*, 53, 116202.

Irimiciuc S. A., Gurlui S., Bulai G., Nica P., Agop M., and Focsa C. 2017b. Langmuir probe investigation of transient plasmas generated by femtosecond laser ablation of several metals: Influence of the target physical properties on the plume dynamics, *Appl. Surf. Sci.*, 417.

Irimiciuc S. A., Bulai G., Gurlui S., and Agop M. 2018b. On the separation of particle flow during pulse laser deposition of heterogeneous materials — A multi-fractal approach, *Powder Tech.*, 339, 273–80.

Irimiciuc S. A., Enescu F., Agop A., and Agop M. 2019a. Lorenz type behaviors in the dynamics of laser produced plasma, *Symmetry*, 1135, 1–13.

Irimiciuc S. A., Gurlui S., and Agop M. 2019b. Particle distribution in transient plasmas generated by Ns — Laser ablation on ternary metallic alloys, *Appl. Phy. B*, 125(10), 1–11.

Kumari S., Kushwaha A., and Khare A. 2012. Spatial distribution of electron temperature and ion density in laser induced ruby (Al 2 O 3:Cr 3+) plasma using langmuir probe, *J. Instrum.*, 7(05), C05017–C05017.

Marine W., Nadezhda N. M., Bulgakova M., Patrone L., and Ozerov I. 2004. Electronic mechanism of ion expulsion under UV nanosecond laser excitation of silicon: Experiment and modeling, *Appl. Phys. A*, 79(4–6).

Merlino R. L. 2007. Understanding langmuir probe current-voltage characteristics, *Am. J. Phys.*, 75(12), 1078–85.

Nica P., Agop M., Gurlui S., and Focsa C. 2010. Oscillatory langmuir probe ion current in laser-produced plasma expansion, *Europhy. Lett.*, 89(6), 65001.

Nica P., Agop M., Gurlui S., Bejinariu C., and Focsa C. 2012. Characterization of aluminum laser produced plasma by target current measurements, *Jpn. J. Appl. Phys.*, 51(10R), 106102.

Nica P., Vizureanu P., Agop M., Gurlui S., Focsa C., Forna N., Ioannou P. D., and Borsos Z. 2009. Experimental and theoretical aspects of aluminum expanding laser plasma, *Jpn. J. Appl. Phys.*, 48, 1–7.

Ojeda G-posada A. 2016. Physical processes in pulsed laser deposition. PhD Thesis, ETH Zurich.

Pompilian O. G., Dascalu G., Mihaila I., Gurlui S., Olivier M., Nemec P., Nazabal V., Cimpoesu N., and Focsa C. 2014. Pulsed laser deposition of rare-earth-doped gallium lanthanum sulphide chalcogenide glass thin films, *Appl. Phys A: Mater. Sci. Process.*, 117(1).

Popa G. and Sirghi L. 2000. *Baiscs of Plasma Physics.* Iasi: Alexandru Ioan Cuza Publisher House.

Salvatore A., Schou J., Lunney J. G., and Phipps C. 2010. Ablation plume dynamics in a background gas, *AIP Conf. Proc.*, 1278, 665.

Schou J. 2009. Physical aspects of the pulsed laser deposition technique: The stoichiometric transfer of material from target to film, *Appl. Surf. Sci.*, 255(10), 5191–98.

Schou J., Toftmann B., and Amoruso S. 2004. Dynamics of a laser-produced silver plume in an oxygen background gas. Proceedings, High-Power Laser Ablation, Vol. 5448, pp. 22–26.

Singh S. C., Fallon C., Hayden P., Mujawar M., Yeates P., and Costello J. T. 2014. Ion flux enhancements and fluctuations in spatially confined laser produced aluminum plasmas, *Phys. Plasmas*, 21(9).

Tang E., Xiang S., Yang M., and Li L. 2012. Sweep langmuir probe and triple probe diagnostics for transient plasma produced by hypervelocity impact, *Plasma Sci. Tech.*, 14(8), 747–53.

Thestrup, B., Toftmann B., Schou J., Doggett B., and Lunney J. G. 2002. Ion dynamics in laser ablation plumes from selected metals at 355 Nm, *Appl. Surf. Sci.*, 197–198, 175–80.

Ursu C. 2010. Caracterisation Par Methodes Optiques et Electriques Du Plasma Produit Par Ablation Laser. Université Lille 1 — Sciences Et Technologies.

Vivien C., Hermann J., Perrone A., and Boulmer-Leborgne C. 1999. A study of molecule formation during laser ablation of graphite in low-pressure ammonia, *J. Phys D: Appl. Phys.*, 32(4), 518–28.

Wen S.-B., Mao X., Greif R., and Russo R. E. 2007. Laser ablation induced vapor plume expansion into a background gas. II, *Exp. Anal. J. Appl. Phys.*, 101(2), 023115.

Yang Q. I., Zhao J., Zhang L., Dolev M., Fried A. D., Marshall A. F., Risbud S. H., and Kapitulnik A. 2014. Pulsed laser deposition of high-quality thin films of the insulating ferromagnet EuS, *Appl. Phys. Lett.*, 104(8).

Chapter 5

Dynamics of Transient Plasmas Generated by Nanosecond Laser Ablation of Metallic Alloys

5.1. Introduction

The dynamics of the ejected particles as a result of a high-power laser and solid matter is not a trivial problem, as it was showcased in several papers (Ojeda *et al.*, 2017; Vitiello *et al.*, 2005). The problem with complex materials, as it is the case with metallic alloys, consists of differences in the physical properties of the composing elements. Phenomena such as heterogenous melting and vaporization (dos Santos Augusto *et al.*, 2017) are commonly reported for nanosecond laser ablation, with dire consequences for applications, such as pulsed laser deposition (PLD). Target material heterogeneity should be reflected in the dynamics of the ejected particles, which is often difficult to observe in industrial applications, such as laser welding, cutting and surface cleaning, but otherwise excellently showcased in applications, such as LIBS or plasma spectroscopy. The amalgam of plasma entities found in a transient plasma generated by laser ablation contains ions, atoms, molecules, electrons and photons. The most often used technique, eloquently showcased by the other groups (Diwakar *et al.*, 2014) or even by our group, is the optical investigation, as it is noninvasive and can differentiate between the contributions of each individual component of the plasma and reflect the complex local and global phenomena reported in recent years.

Understanding laser-based technologies and the interaction between a high-energy laser beam and metallic alloys is now relevant for a wide range of applications with fast feedback and accurate predictions on the behavior of physical processes. The dual approach of experimental investigations and theoretical modeling has proven to be a successful method for understanding the dynamics of multi-element fluids (Qiu *et al.*, 2012; Zhou, 2017) and, as it was showcased recently by our group, complex laser-produced plasmas (LPPs) (Irimiciuc *et al.*, 2018). The study presented in this chapter continues our previous endeavors with respect to the stoichiometric transfer and plasma chemistry in the case of the laser ablation of complex alloys. We discuss here the ejection of metallic particles as a result of laser ablation of ternary alloys from both experimental and theoretical points of view. To understand the dynamics of ejected metallic ions, we implemented optical emission spectroscopy (OES) and ICCD fast camera imaging to attain global and local information about their kinetic and thermal energy and their spatial distribution within the ablated cloud. From a theoretical perspective, we expanded on our model from (Irimiciuc *et al.*, 2018) and focused on understanding, under a fractal movement paradigm, the effects of the plasma temperature and ion mass on the spatial distribution of complex alloy plasmas.

Understanding the flow of multi-phase fluids containing mixtures has become one of the main areas where numerical modeling can help aid already existing technologies. For techniques, such as spray pyrolysis (Mardare *et al.*, 2014), electrospray deposition (Dizdar *et al.*, 2018) or even PLD, it is important to know how the dynamics of individual particles can affect the final product. The presence of *heavier* particles in the flow of a particular fluid can strongly affect the outcome of the flow, either when relating to pulverized oil (Love *et al.*, 2014) or in a general case (Zhou, 2017) where the possibility of turbulent flow is discussed. Investigations on the individual ejected particles can be used for the study of multi-phase fluids and further envelop the control of such complex systems, as is the case for industrial applications. PLD has been known for some time now as a source for both generation of complex thin films (Lin *et al.*, 2004) or nanoparticles (Ausanio *et al.*, 2006) and clusters in a simultaneous manner, with reports of turbulent plasma flow for configurations with strongly heterogeneous particle distributions (Borowitz *et al.*, 1987). As such, theoretical modeling and numerical simulations become powerful tools for understanding the behavior of real fluids under specific conditions. However, the transfer from theory to application is rather difficult as most models work with an idealized representation of real complex fluids (Batchelor, 1999). The dynamics describing the flow of complex fluids (plasmas, oils, multi-component mixtures) are influenced and, to some extent, defined by the multiple interactions between its structural units (i.e. different particles (Lakes 2009), fluid–solid (Verloop and Heertjes, 1970) or gas–solid (Geldart, 1973) mixtures, organic fluids (Cushing *et al.*, 2004), etc.), and their evolution cannot be predicted simply by investigating the individual behavior. Actually, their global evolution can be determined by using an approach that considers the individual structural units and their interactions with one another. The most significant properties of the complex fluids becoming emergent are self-organization and adaptability.

Classical models used to investigate the dynamics of complex fluids are based on an unjustified assumption of differentiability of the physical variables (e.g. density, momentum, energy (Nedeff *et al.*, 2015a; Kelessidis and Mpandelis, 2004; Monaghan, 1992; Zhang *et al.*, 2009)) and the processes which they define. The usefulness of such methods can be understood sequentially in space–time domains for which the differentiability is not broken. However, this (differential) approach sometimes fails when it is confronted by the reality of a complex fluid (i.e. the flow during spray pyrolysis or plasma plume expansion in PLD). For a better representation of all the interactions at both the local and global scales, we need to introduce in an explicit manner the dependence on the scale resolution. This translates into a new physical system where all the dynamic variables that were dependent only on space and time in a classical way will now be dependent on the scale resolution. This approach can be further simplified, and instead of working with nondifferential functions, which can be rather difficult, we will just use different approximations of nondifferential mathematical functions derived through their averaging at various scale resolutions.

A major consequence of this approximation is that any dynamic variable acts as the limit of specific function families, which are nondifferentiable for a null-scale resolution.

5.2. Transient plasmas generated by nanosecond laser ablation on ternary metallic alloys

5.2.1. *Space-and time-resolved optical emission spectroscopy*

When investigating the ejected cloud of particles, the *ideal* investigation technique should be noninvasive and offer global and local information about the plasma components. Such a technique can be considered a combination of ICCD fast camera imaging and space- and time-resolved OES, which is used consistently by our group (Irimiciuc *et al.* 2017b, 2018), and it was also validated by a significant number of papers (Ngom *et al.*, 2016; Singh *et al.*, 2017). Our approach was a global–local one, meaning that first we recorded the overall emission of the LPP at various time delays with respect to the laser beam. The result for the Fe–Mn–Si plasma is represented in Figure 5.1, where we present selected images across the evolution of the plasma in a 2 μs time lapse.

We observe that the plasma has a quasi-spherical shape and increases its volume as the plasma evolves. The expansion velocity was estimated using the technique presented in other previous papers (Ojeda *et al.*, 2017; Cherhrghani *et al.*, 2013), where the effect of multi-element composition on the ablation process is discussed. When performing cross-section on the recorded images

Figure 5.1. Global evolution of the Fe–Mn–Si LPP.

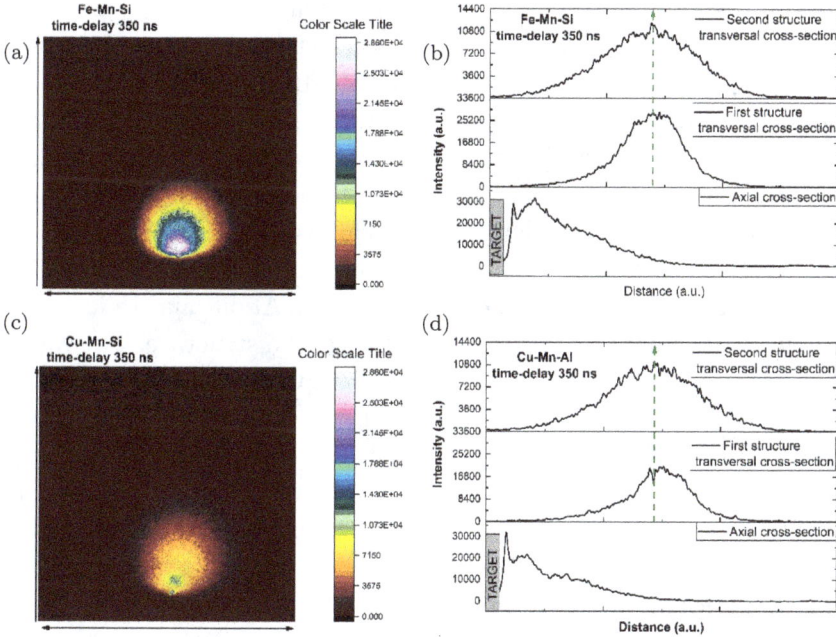

Figure 5.2. ICCD images of LPP generated on Fe–Mn–Si (a) and Cu–Mn–Al (d) samples and their respective cross-sections (b and c).

(Figure 5.2(a)–(d)) in the axial and transversal directions, we notice different behaviors across the two directions. In our previous paper (Irimiciuc *et al.*, 2018), we noticed a splitting in the transversal axis; however, in this case, we do not notice this phenomenon. These differences can be explained by the overall lower fractality degree of our physical system as compared to our previous stainless-steel LPP. We note here that the lateral splitting is seen as a fingerprint of an elevated fractality degree (Irimiciuc *et al.*, 2018).

The cross-section performed across the main expansion axis (axial cross-section) reveals a splitting of the plasma plume into three distinct structures. In general, the literature presents a specific nomenclature for these structures: the *first one*, also named the *fast one*, is generated by electrostatic interactions; the *second one* (*slower structure*) is generated by thermal mechanisms; and the third one contains mainly molecules, clusters or nanoparticles. Their presence has previously been reported by our group and extensively discussed in conjecture with the multiple-ejection mechanism and the fractality of the LPP (Irimiciuc *et al.*, 2014, 2017, 2018). However, our focus will not be on this third structure, as the main optical signatures, seen through our experimental methods, are given by the dynamics of simpler plasma entities, such as atoms or ions. The velocities of the first two structures were determined as follows: for the case of the Fe–Mn–Si plasma, 20 km/s for the first structure and 11.2 km/s for the second one, while for the case of the Cu–Mn–Al plasma, 15 km/s for the first structure and 7.45 km/s for the second structure. The values are in good agreement with the usual reports from the literature (Ojeda *et al.*, 2017; Cherhrghani *et al.*, 2013). The obtained values are strongly related to the differences between the melting points for each material and the overall mass of the cloud, with significant variance in the properties of the component directly affecting the ablation process and the subsequent evolution.

We notice a significant difference in the overall emission and shape of the LPP generated on the two alloys. The global emission is notably larger for the Fe–Mn–Si plasma and with less inner structuring, while for Cu–Mn–Al, the global emission is reduced and presents more pronounced structuring. These differences are a reflection of the uniformity in the energy distribution of the excitation process as opposed to ionization or other types of interactions. The uniform aspect of the Fe–Mn–Si plasma can be attributed to the fact that, out of the two ternary alloys, the composing elements, in this case, present relatively close melting points, which leads to a uniform and homogeneous ablation. On the other hand, for the Cu–Mn–Al plasma, the significant differences between the physical properties of Al and Mn or Cu could lead to a more heterogenous ablation process. These statements will be further verified through the space- and time-resolved OES. We would also like to note that the fractality of the LPPs will also be affected by the inner energy of the plasma and its distribution on the composing entities (Irimiciuc *et al.*, 2018a, 2018b). We anticipate here another type of analysis (fractal analysis), which we will further use in this study, that could offer valuable information about the LPPs.

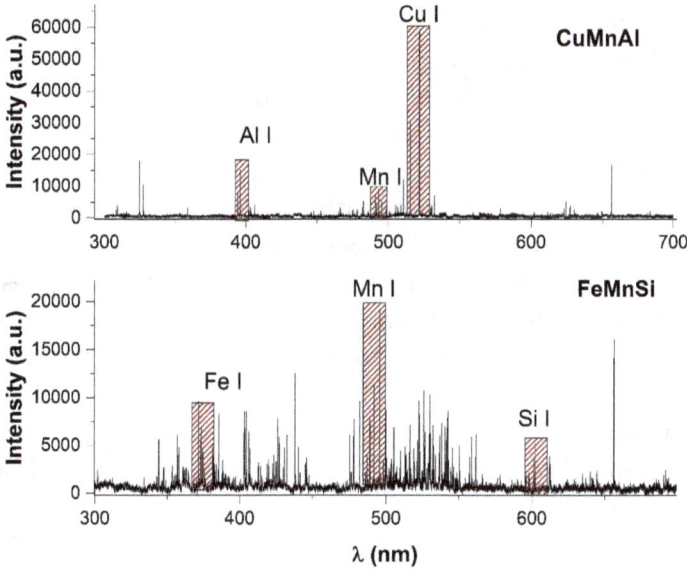

Figure 5.3. Optical emission spectra recorded for a gate width of 1 μs along the main expansion axis of the plasma plume.

The main advantage of the OES approach is in dealing with the information from the LPP attributed to individual plasma species. As such, we are able to differentiate between the excitation energy and the frequency of appearance of certain species. In Figure 5.3, we represent the global spectra recorded over 2 μs along the main expansion axis for both LPPs. We can identify, by using the specialized databases, the emission lines characteristic of all the components of the alloys in both atomic and ionic states (Al, Cu, Mn, Si and Fe).

Using the methodology described by Cristoforetti *et al.* (2006) and Irimiciuc *et al.* (2016), we can extract information regarding the energy lost during excitation processes, electron density and the presence or absence of the local thermodynamic equilibrium (LTE). The excitation temperature was determined by using the Boltzmann plot method (Singh and Thakur, 2007). We chose to implement it individually, per each species, in order to observe subtle changes in the excitation energy induced by differences in the collision frequency between different species of atoms and electrons. The values found for each element are as follows: Cu (0.74 eV), Al (0.71 eV) and Mn (1.47 eV) for the Cu–Mn–Al plasma; Fe (1.72 eV), Mn (1.02 eV) and Si (0.4 eV) for the Fe–Mn–Si. We notice some slight differences in the excitation temperatures of the composing elements. In the case of the Cu–Mn–Al plasma, the Cu and Al species are described by similar excitation energies, with the Mn species showcasing an elevated excitation energy. Furthermore, in the case of Fe–Mn–Si, these differences are clearer, with each element being described by an individual excitation temperature. This heterogeneity should not be seen as a deviation from the LTE, as each individual species respects the LTE McWhirter criterion (Hahn and Omenetto, 2010). In our previous paper, the differences between the individual excitation energies were attributed to the differences in the fractality of each species (Irimiciuc *et al.*, 2018b). Here, we attempt to elevate that vision for multi-element plasmas and understand how different fractality degrees for each plasma entity can affect the spatial distribution of the respective species. We would like to stress that the heterogeneity in the values of the excitation temperature is given by the different spatial distributions of the elements at any given time within the plasma plume, meaning that each species will interact with particles with certain energies (e.g. the species closest to the target will interact with high-energy high-density plasma particles, while those with a wider distribution will reflect the energy and density distributions of the plasma plume).

In Figure 5.4, we represent the spatial distribution of Fe and Mn atoms from the Fe–Mn–Si plasma, showcasing the difference between the two elements. We would like to note that Si was not taken into account, as the emission line intensity for its species was significantly lower. We notice that the Fe atoms have a dual peak distribution, while the Mn one presents only a single peak distribution. This means that Fe atoms are excited throughout the plasma volume even at a longer distance, where the electron density is significantly lower. This result also explains the elevated T_{ex} presented in the previous paragraph and is in line with the multiple-structure scenario seen through ICCD fast camera imaging. By representing the maximum of luminous intensity for each emission line as a function of time

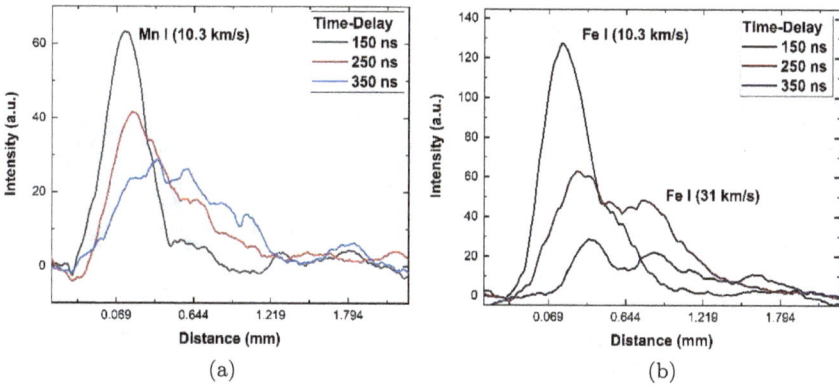

Figure 5.4. Axial distribution of Fe (a) and Mn (b) atomic emission at various time delays.

(Geohegan *et al.*, 1998), we determined the expansion velocities of the individual elements, with 31 km/s for the first peak of Fe I and 10.3 km/s for the second one, while for the Mn I, the velocity was 18 km/s. The values of the first peak of Fe I and the velocity of the Mn atoms are in line with the values of the second plasma structure, while the value of the second group of Fe atoms is similar to the value determined for the first plasma structure. These results reveal that the two-plasma structure has uniformly distributed atoms and ions among them, with the fast structure having a slight depletion of Mn.

We can generalize the discussion made above for both the investigated plasmas as the irradiation and expansion conditions are identical (background pressure and laser fluence). The results are presented in Figure 5.5 (left), where we can observe after 150 ns the spatial distribution of Fe and Mn in the Fe–Mn–Si plasma and Cu and Al in the Cu–Mn–Al plasma. We notice that for lighter elements, we obtain a narrow spatial distribution, while the *heavier* ones (Cu and Fe) have a wider distribution. These differences can be seen as a separation of the composing elements based on their physical properties. The separation was previously discussed by our group (Irimiciuc *et al.*, 2018), where the fractality of the components played a significant role, based on which the spatial distributions of different elements are reflective of the elevated degree of fractality. Lighter

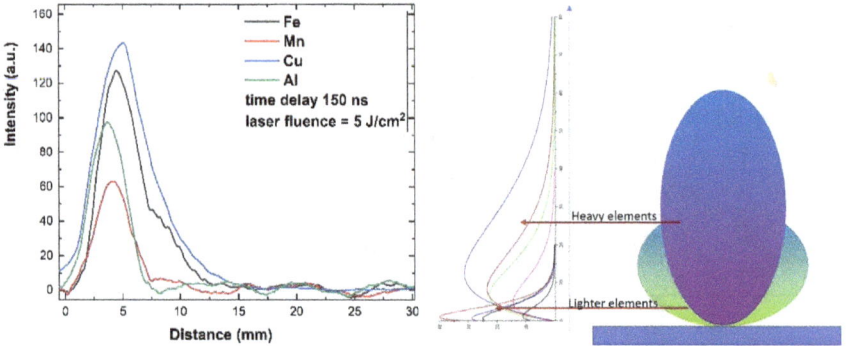

Figure 5.5. Axial distribution of the main elements in the alloys as seen through OES measurements at a time delay of 150 ns (left) and a schematic representation of the particle distribution within the plasma volume (right).

elements will have a higher collision rate and thus a higher fractality degree, whereas the heavier ones are described by a lower fractality degree (lower collision rate). This difference in the fractality of the plasma entities will give us different spatial distributions for each element.

However, given our optical configuration setup, lighter elements strongly scattered during expansion will appear to have a narrower distribution at relatively short distances, while heavier particles will have a broader distribution, most likely covering the whole plasma plume. Translating these results into the real dynamic of a 3D plasma, lighter elements are scattered toward the edge of the plume, and the heavier ones constitute the core of the LPP. For industrial applications, such as PLD, the result is of great interest, as the inner structure of the plasma plumes lacks stoichiometry and uniformity. These properties could lead to the nonstoichiometric transfer of complex metallic alloys and change the physical properties of the final product. Furthermore, the diagnostic system used here allowed us to capture the complex nature of the plasma and present some meaning behind it. We attempt to further unravel more information about the relation between the fractality of specific elements and their spatial distributions within the plasma volume in the following section.

5.2.2. *Multifractal analysis*

The fractal analysis approach to understanding the dynamics of complex physical systems was shown over the years to provide with some of the most promising results toward understanding multiparticle flow in fluids (Merches and Agop, 2015; Agop and Merches, 2019) or plasmas (Irimiciuc *et al.*, 2014, 2017, 2018).

A deterministic approach does not necessarily involve an ordinary (periodic movement, self-structuring, etc.) or periodic behavior of LPPs. In the fractal analysis on which the image of the LPP was developed theoretically, the nonlinear periodicity appears automatically as a quality of its dynamics. The development of nonlinear analysis and the discovery of a series of laws that govern the chaos showcased not that the reductionist analysis method, on which the entirety of plasma physics was based, has a limited applicability. Also, such an approach presented that the uncontrolled predictability is not a property of laser ablation plasmas but a natural consequence of their simplification through linear analysis. It follows that the nonlinearity and chaos present common behaviors, meaning a universality of the mathematical laws that govern the transient plasma dynamics.

For a laser ablation plasma, the nonlinearity and the chaoticity have a dual applicability, being both structural and functional, with the interactions between the so-called plasma entities (structural components, such as electrons, ions, atoms and photons) determine reciprocal conditioning: micro–macro, local–global, individual–group, etc. In such a case, the universality of the laws describing the laser ablation plasma dynamics becomes obvious, and it must be reflected by the mathematical procedures which are utilized. Basically, it often makes more and more use of the "holographic implementation" in the description of plasma dynamics. Usually, the theoretical models used to describe the ablation plasma dynamics are based on a differentiable variable assumption. Most of the notable results of the differentiable models must be understood sequentially, where the integrability and differentiability still apply. The differentiable mathematical procedures are limiting our understanding of more complex physical phenomena, such as the expansion of an LPP, which implies various nonlinear behaviors, chaotic movement and self-structuring. In order to accurately describe the LPP dynamics and still remain tributary to differentiable and integral mathematics, we must explicitly introduce the scale resolution. The scale resolution will be integrated into the expression of the physical variable, which describes the LPP, and implicitly in the fundamental equations, which govern these dynamics. This means that any physical variable becomes dependent on both the spatial and temporal coordinates and the scale resolution. In other words, instead of using physical variables described by a nondifferentiable mathematical function, we will use different approximations of this mathematical function obtained through its averaging at various scale resolutions. As a consequence, the physical variables used to describe the LLP dynamics will act as a limit of the functions family, which are nondifferentiable for a null-scale resolution and differentiable for a non-null-scale resolution.

This method of describing LPP dynamics implies the development of new geometric structures (Cristescu, 2008; Mandelbrot, 1993) as well as new physical theories, for which the movement laws that are invariant to spatiotemporal transformation are integrated on scale laws invariant to scale resolution transformations. In our opinion, such geometric structures can be produced by the fractal–multifractal theory of movement, either in the form of scale relativity theory (SRT) in the fractal dimension $D_F = 2$ (Nottale, 2011) or in the form of SRT in an arbitrary constant fractal dimension (Merches and Agop, 2015; Agop and Paun, 2017). In either of the two cases, the *holographic implementation* of specific dynamics of LPP implies the substitution of dynamics with restrictions in an Euclidian space, with dynamics free of any restriction in a multifractal space. Thus, we make use of only the movement of the plasma particles on continuous and nondifferentiable curves in a multifractal space (Nottale, 2011).

In the following, we analyze some specific dynamics of a transient plasma generated by laser ablation, therefore postulating that the plasma particles are moving on multifractal curves. The mathematical procedure implies the usage of the following set of multifractal hydrodynamic equations (Merches and Agop, 2015; Agop and Paun, 2017).

In such a context, let us consider the density current:

$$\Im(x,t,dt) = \rho(x,t,dt)V(x,t,dt)\sum = \frac{\Sigma}{\pi^{1/2}} \frac{V_0\alpha^2 + \frac{4\lambda^2(dt)^{\frac{4}{F(\sigma)}-2}}{\alpha^2}xt}{\left[\alpha^2 + \frac{4\lambda^2(dt)^{\frac{4}{F(\sigma)}-2}}{\alpha^2}t^2\right]^{3/2}}$$

$$\times \exp\left[-\frac{(x-V_0t)^2}{\alpha^2 + \frac{4\lambda^2(dt)^{\frac{4}{F(\sigma)}-2}}{\alpha^2}t^2}\right], \tag{5.1}$$

where Σ is a surface which $\Im$ crosses and the other parameters have the meaning given in Chapter 3.

In the aforementioned conditions, $\Im$ is invariant with respect to the coordinate transformation group and to the scale resolution transformation group. Since these two groups are isomorphs, between them, we can unravel various isometries, such as: compactizations of the spatial and temporal coordinates, compactization of the scale resolutions, and compactizations of the spatiotemporal coordinates and scale resolutions. Following this, we can perform a compactization between the temporal coordinate and the scale resolution, which is given by the relation

$$\varepsilon = \frac{E}{m_0} = 2\lambda(dt)^{\frac{2}{F(\sigma)}-1}v, \quad v = \frac{1}{t}, \tag{5.2}$$

where corresponds to the specific energy of the ablation plasma entities. Once admitting such an isometry, by means of the substitutions

$$I = \frac{\Im\pi^{1/2}\alpha}{V_0\Sigma}, \quad \xi = \frac{x}{\alpha}, \quad u = \frac{\varepsilon}{\varepsilon_0}, \quad \varepsilon_0 = \frac{2\lambda V_0(dt)^{\frac{2}{F(\sigma)}-1}}{\alpha},$$

$$\mu = \frac{2\lambda(dt)^{\frac{2}{F(\sigma)}-1}}{\alpha V_0}, \tag{5.3}$$

(5.1) takes the more simplified non-dimensional form

$$I = \frac{1+\mu^2\frac{\xi}{u}}{\left(1+\mu^2\frac{\xi}{u}\right)^{3/2}}\exp\left[-\frac{\left(\xi-\frac{1}{u}\right)^2}{1+\left(\frac{\mu}{u}\right)^2}\right]. \tag{5.4}$$

In (5.3) and (5.4), I corresponds to the normalized state intensity, ξ to the normalized spatial coordinate, u to the normalized specific energy of the ablation plasma entities and μ to the normalized multifractalization degree. Since the specific energy ε and the reference energy ε_0 can be expressed as

$$\varepsilon \approx \frac{T}{M}, \quad \varepsilon_0 \approx \frac{T_0}{M_0}, \tag{5.5}$$

where T and T_0 are the specific temperatures and M and M_0 are the specific masses. We can further note

$$\tau = \frac{T}{T_0}, \quad \theta = \frac{M}{M_0}. \tag{5.6}$$

so that (5.1) becomes

$$I = \frac{1 + \mu^2 \xi \frac{\theta}{\tau}}{\left(1 + \left(\mu\frac{\theta}{\tau}\right)^2\right)^{3/2}} \exp\left[-\frac{\left(\xi - \frac{\theta}{\tau}\right)^2}{1 + \left(\mu\frac{\theta}{\tau}\right)^2}\right]. \tag{5.7}$$

The fundamental behavior of transient plasmas generated by laser ablation can be assimilated with a nondifferentiable medium, and the fractality degree is given by the elementary processes induced by a collision, such as excitation, ionization or recombination (for other details, see Irimiciuc *et al.* (2018a, 2018b, 2014a, 2014b, 2014c). In such a conjecture, (5.1) defines not only the normalized state intensity but also a measure of the spectral emission of each plasma component, a situation in which its various distributions (spatial or mass type) specified through our mathematical model can be correlated with our data.

The results are presented in Figure 5.6(a) and 5.6(b), where one can observe that particles with a fractality of $\mu < 1$ are described by a narrow distribution centered around small values of ξ, while for a fractality of $\mu > 1$, the distribution is larger and is centered around values that are one order of magnitude higher than the high fractality ones. Thus, we formulate an image of the LPP in a fractal mathematical formalism, which has a *core* of entities with a low fractality and a relatively low plasma temperature as well as a *shell* of high energetic particles described by a higher fractality degree.

To compare our results and find how they are related to the classical view of the LPP, we have performed a simulation of the plasma emission distribution over the mass of the particles for a

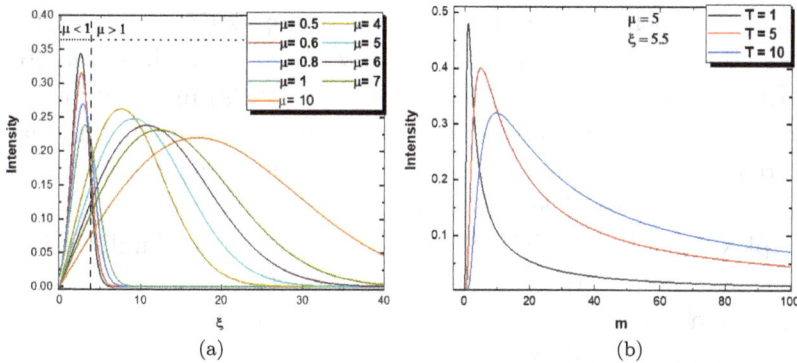

Figure 5.6. Spatial dependence of the simulated optical emission of plasma entities with various fractal degrees (a) and mass distribution of the optical emission for various plasma temperature (b).

plasma with an overall μ factor of 5 at an arbitrary distance ($\xi = 5.5$). We observe that entities with a lower mass have a higher relative emission for a given temperature, and with an increase in the plasma temperature, the emission of heavier elements will increase as well. These results are in good agreement with our previous results from the works of Irimiciuc *et al.* (2018a, 2018b, 2014a, 2014b, 2014c), where we assimilate the plasma temperature with the inner fractal energy of the plasma. The implications of these results can be directly implemented for industrial technologies: For relatively low plasma temperatures, the distribution is strongly heterogenous, and it facilitates only particles with a high fractalization degree, possibly leading to a nonstoichiometric transfer in the case of PLD, and the lighter elements are predominantly in the outer regions of the plume, while the heavier ones are mainly part of the core.

5.3. A multifractal theoretical approach to understanding the separation of particle flow during the PLD of multicomponent alloys

5.3.1. *Complex fluids flow: A nondifferential perspective*

This peculiar approach to the study of complex fluid dynamics has been successfully used for the analysis of various representations of real fluids, such as blood (Merches and Agop, 2015), complex polymers (Bacaita *et al.*, 2016) or even discharge (Dimitriu *et al.*, 2015) or LPPs (Irimiciuc *et al.*, 2017; Agop *et al.*, 2009) in the fluid model representation. Most of the powerful results of the nondifferentiable approach consist in the interpretation of the influences induced by the composing particles (we call them the *structural units* of our complex fluid) on the global dynamics of the complex fluid. Basically, we focus on how the whole system reflects the interconnectivity and interactions of all its building blocks (i.e. how is the flow of a real fluid affected by the interactions of its structural units – atoms, molecules, clusters, nanoparticles, etc.). A major problem in most of their applications is the separation and structuring of the *real* fluid (in the representation of vapor, plasma, polymer or fluids) during various technological processes. The model attempts to showcase that the structuring processes, regardless of the type of fluid, are manifestations of the same physical forces at different scale resolutions.

The particularities of the model have already been discussed in other papers published by our group (Irimiciu e*t al.*, 2017a). Briefly, we consider that the movement of the complex fluid components (structural units) is defined up to some limits by continuous but nondifferential curves. This allows us to further project the properties of real fluids in a fractal matrix, thus simplifying the dynamics of the individual particles or molecules by assimilating them with their respective fractal geodesics (i.e. their trajectories). Thus, for large time scales with respect to the inverse of the maxim (Cristescu, 2008) Lyapunov exponent, the deterministic trajectories are replaced by families of potential trajectories (i.e. fractal geodesics), and the concept of defined positions are replaced by that of probability densities.

In such a context, in agreement with the results from Merches and Agop (2015) and Agop and Paun (2017), at a differentiable resolution scale, the ablation plasma dynamics are driven by the specific fractal force

$$F_F^i = \left[u_F^l + \frac{1}{4}(dt)^{\left(\frac{2}{D_F}\right)-1} D^{kl} \partial_k \right] \partial_l u_F^i. \tag{5.8}$$

The presence of this specific fractal force in an explicit manner could be responsible for the separation of the complex fluid into each component by introducing a special velocity field. To this end, we admit the functionality of our differential system of equations:

$$F_F^i = \left[u_F^l + \frac{1}{4}(dt)^{\left(\frac{2}{D_F}\right)-1}D^{kl}\partial_k\right]\partial_l u_F^i = 0, \qquad (5.9)$$

$$\partial_l u_F^l = 0. \qquad (5.10)$$

The first equation specifies the fact that for a differential scale resolution, the fractal force becomes null, while the second one represents the state density conservation law at a nondifferentiable scale resolution.

In general, it is difficult to obtain an analytic solution for our system of equations, taking into account its nonlinear nature (through the fractal convection $u_F^l \partial_l u_F^i$ and the fractal-type dissipation $D^{kl}\partial_l \partial_k u_F^i$) and the fact that the fractalization type, given by the fractal type tensor D^{kl}, remains unknown in these representations.

For further development of our model and its implementation for the study of *real* or *abstract* fluids, we define the flow of a three-dimensional (3D) fluid with a revolution symmetry around the z-axis, and investigate its dynamics through the two-dimensional (2D) projection of the fluid in the (x, y) plane.

By considering only the symmetry plane (x, y), the system (5.9) and (5.10) becomes

$$u_{F_x}\frac{\partial u_{F_x}}{\partial x} + u_{Fx}\frac{\partial u_{F_x}}{\partial y} = \frac{1}{4}(dt)^{(2/D_F)-1}D^{yy}\frac{\partial^2 u_{F_x}}{\partial y^2}, \qquad (5.11)$$

$$\frac{\partial u_{F_x}}{\partial x} + \frac{\partial u_{F_y}}{\partial y} = 0. \qquad (5.12)$$

Let us solve the equation system (5.11) and (5.12) by imposing the following conditions:

$$\lim_{y\to 0} u_{F_y}(x, y) = 0, \lim_{y\to 0}\frac{\partial u_{F_x}}{\partial y} = 0, \lim_{y\to\infty} u_{F_x}(x, y) = 0, \qquad (5.13)$$

$$\Theta = \rho \int_{-\infty}^{+\infty} u_x^2 dy = \text{const.},$$

with

$$D^{yy} = a\exp(i\theta). \qquad (5.14)$$

We would also like to note that the presence of the complex phase can lead to a hidden temporal evolution of the system. It is known that the variation of a complex phase defines implicitly a time dependence which would make our system capable of depicting both space and time evolutions. Thus, the choice of D^{yy} can lead to the possibility of both a spatial and a temporal study of the dynamics of our system.

The solution to equations (5.11) and (5.12) in their outmost general form with normalized quantities

$$X = \frac{x}{x_0}, \quad Y = \frac{y}{y_0}, \quad U = u_{F_x} \frac{4y_0^2}{x_0 a}, \quad V = u_{F_y} \frac{4y_0^2}{x_0 a},$$

$$\frac{\left(\frac{\Phi_0}{6\rho}\right)^{\frac{1}{3}}}{\left(\frac{a}{4}\right)^{2/3}} = \frac{x_0^{2/3}}{y_0}, \quad \mu = \left(dt()^{\left(\frac{D_F}{2}\right)^{-1}}\right) \tag{5.15}$$

is given according to the method given by Carafoli and Constantinescu (1984b):

$$U(X,Y) = \frac{\frac{3}{2}}{[\mu X]^{\frac{1}{3}} \exp\left(\frac{i\theta}{3}\right)} \cdot \operatorname{sech}^2 \left\{ \frac{\frac{1}{2}Y}{[\mu X]^{\frac{2}{3}} \exp\left(\frac{2i\theta}{3}\right)} \right\}, \tag{5.16}$$

$$V(X,Y) = \frac{\left(\frac{9}{2}\right)^{\frac{2}{3}}}{[\mu X]^{\frac{1}{3}} \exp\left(\frac{i\theta}{3}\right)} \left\{ \left[\frac{Y}{\mu X^{\frac{2}{3}} \exp\left(\frac{2i\theta}{3}\right)} \right] \right.$$

$$\cdot \sec h^2 \left[\frac{\frac{1}{2}Y}{[\mu X]^{\frac{2}{3}} \exp\left(\frac{2i\theta}{3}\right)} \right] - \tanh \left[\frac{\frac{1}{2}Y}{[\mu X]^{\frac{2}{3}} \exp\left(\frac{2i\theta}{3}\right)} \right] \right\}. \tag{5.17}$$

To verify the validity of such an unusual approach, we obtained 3D (Figure 5.7) modeling representations depicting the flow of a complex fluid based on the solution given by our system of equations. The fluid is defined in the framework of our model as a mixture of various particles with different physical properties. As a result, parameters, such as the complex phase, fractal dimension or specific length (x_0, y_0), will incorporate within their values the unique properties of each of the component. Figure 5.7 showcases the separation of the flow for different values of the complex phase, leading to the appearance of preferential *flow lines* for $\Theta > 1.5$.

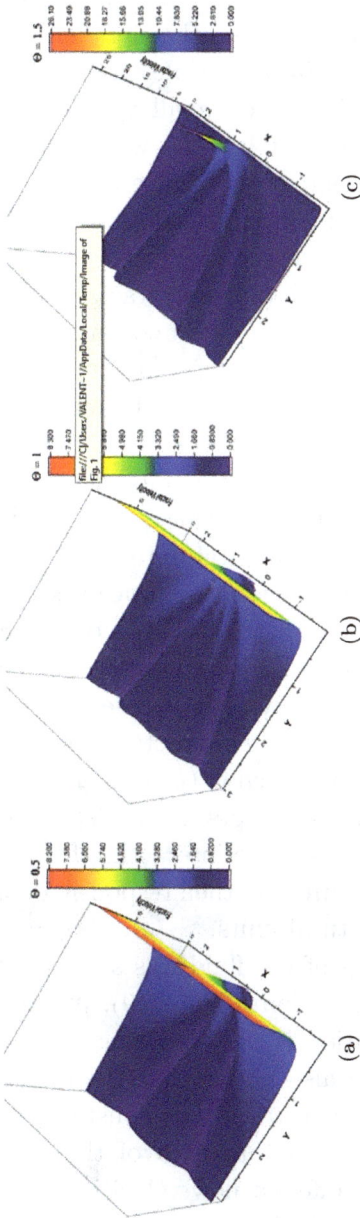

Figure 5.7. A 3D representation of the total fractal velocity field of a multifractal fluid for various complex phases: 0.5 (a), 1 (b) and 1.5 (c).

Figure 5.8 shows the 2D representations depicting various flow scenarios with respect to the structure of the fluid, starting from a pure uni-particle fluid toward a multi-component fluid. We observe that there is a segregation into multiple structures in all expansion directions (across X and Y). For small values of θ, which will be our *control* parameter in the following, we define a fluid with only one component. As we can see in Figure 5.8(a), there is only one fluid structure along the main expansion axis. With an increase in this parameter and subsequent changes in the homogeneity of the structural units of the fluid (i.e. our fractal fluid becomes more heterogeneous in terms of both size and energy of the structural units), we observe the formation of two symmetrically positioned secondary structures. These areas contain mainly structural units with a small physical volume and low kinetic energy. A further increase in the heterogeneity of the fluid leads to the formation of symmetrically situated fluid *structures*, each defining a family of physical properties for the structural units.

We observe that the self-structuring process is gradual. For values of $\theta = 0.4 - 1$, we obtain the three main structures which are followed by a subsequent internal structuring visible later on for $\theta > 1$. We would like to point out that this is a reversible transition as the distribution often returns to a tree structure configuration. This can be seen as an modulated behavior (*breathing modes of the fractal fluid*) for our fluid, which attempts a complete transition toward a completely separate flow, but the interacting fractal forces between the individual fluid structures are then responsible for the unification of the fluid and its structural units.

The multi-structuring of the fluid was showcased by performing cross-sections in the X direction (Figure 5.9). We observed that in the X direction, the separation is more striking in the incipient states of flow (expansion). We also notice that the distance between the maxima of the three structures is not constant during expansion, leading us to the conclusion that each of the newly created fluid structures is defined by different flow velocities. Some investigations were performed by implementing the same data treatment for the Y

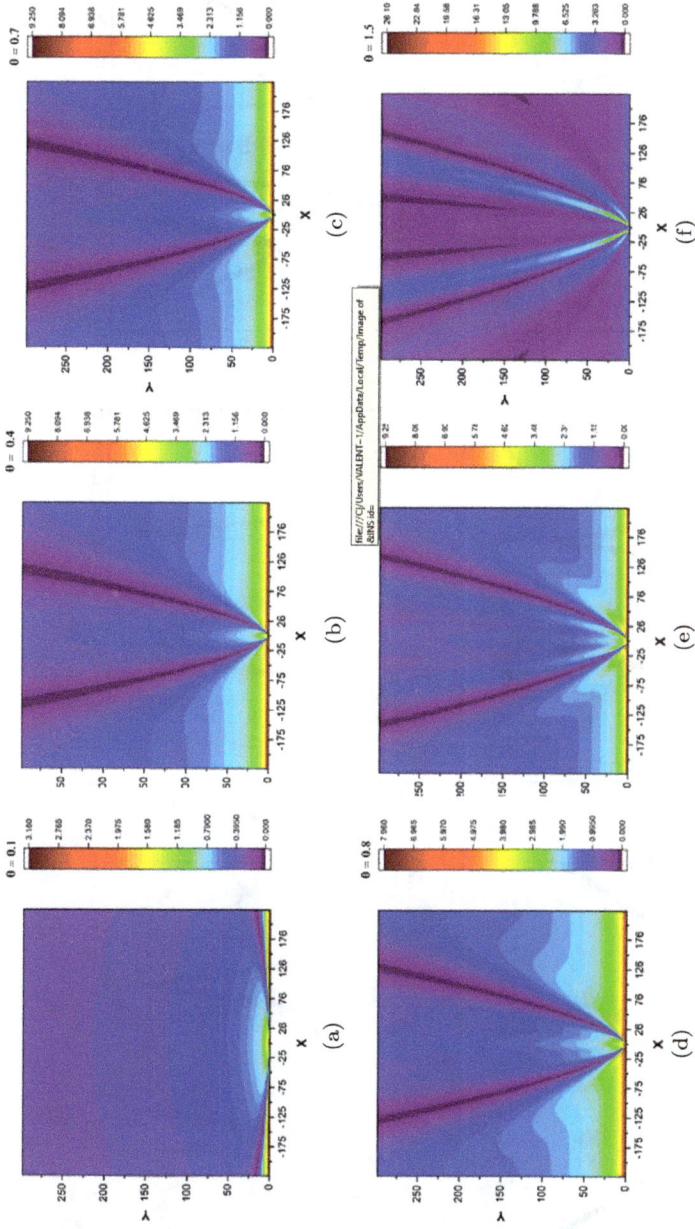

Figure 5.8. Total fractal velocity field evolution in the two main directions (X,Y) for a multifractal system with θ values of 0.1 (a), 0.4 (b), 0.7 (c), 0.8 (d), 1 (e) and 1.5 (f).

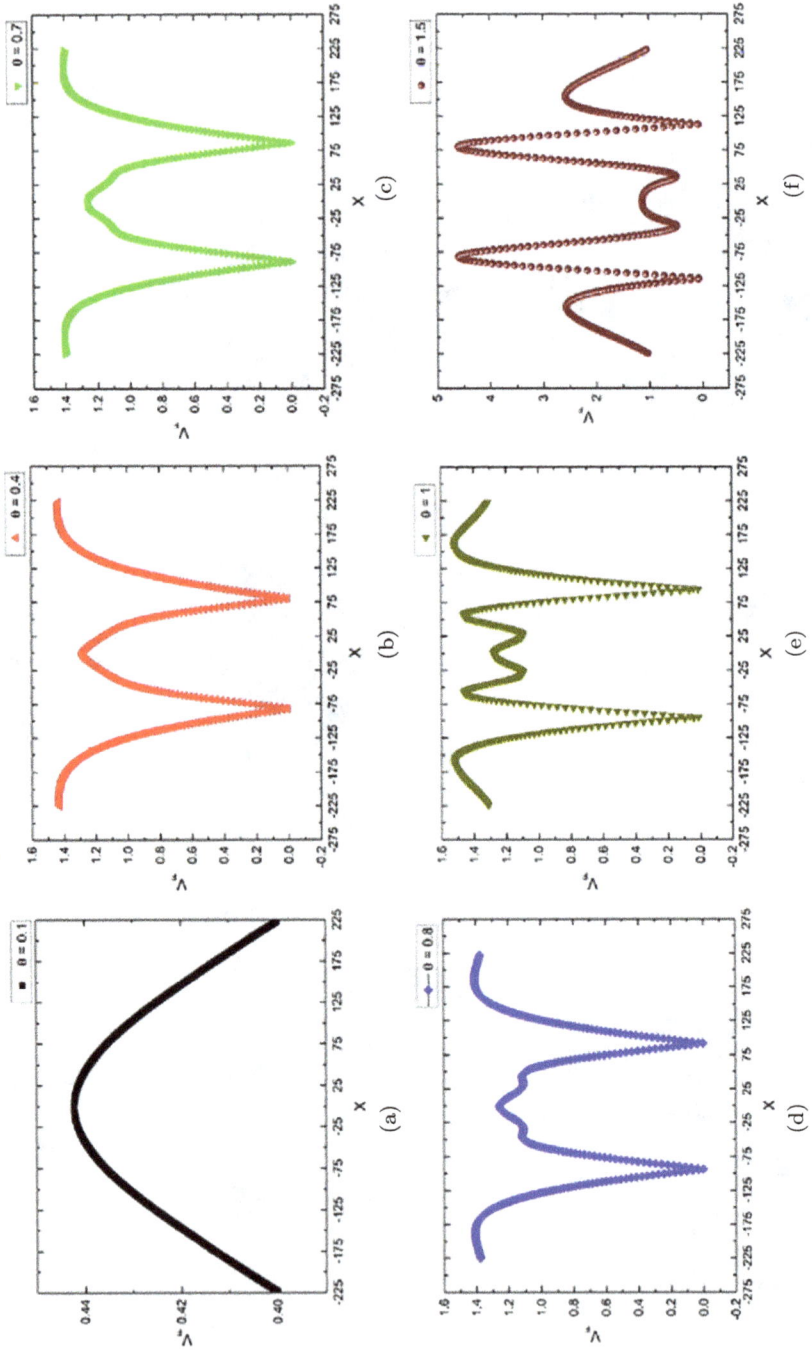

Figure 5.9. Transversal cross-sections of a fractal fluid's global velocity field.

direction. For the cross-section on the Y-axis (at $X = 0$), we observe another, less clear, separation. This is understandable, as the structuring phenomenon of the fluid is not restricted to a particular flow axis, being observed in all directions. Also, the fractality of the system, defined here through θ and μ, is directly related to the trajectory of the fluid *particles*. As such, for a more complex fluid, the intrinsic dynamic of the particles will lead to a separation on the main expansion axis (at $X = 0$).

Our theoretical approach manages to describe the structuring of the complex fluid. However, this is still an abstract view of a real problem in technology. In order to validate both the conceptual and mathematical approaches, we choose to perform experimental investigations of LPPs under similar conditions to those used for PLDs. This is a suitable physical phenomenon, as in recent years, various groups have shown (Sloyan *et al.*, 2009; O'Mahony *et al.*, 2007; Canulescu *et al.*, 2017, 2009a) that in the case of complex plasmas, there is separation of the particles during expansion based on their properties, which could affect the quality and structure of the deposited film. The study was performed by means of space- and time-resolved OES and ICCD fast camera photography, implemented for the capture of local and global dynamics of a nanosecond LPP on a stainless-steel target.

5.3.2. *Empirical confirmation of the multifractal paradigm: Dynamics of a stainless-steel laser-produced plasma*

5.3.2.1. *Global aspects: ICCD fast camera imaging*

In order to investigate how the presence of multiple components in a metallic target affects their dynamic post-laser–matter interaction and subsequent ejection, two types of investigations were implemented. The global ICCD imaging technique was done with the purpose of studying if the dynamic of the plume confirmed the fractal paradigm presented in the previous section (namely, its heterogeneity in the expansion velocity field reflected in the individual energy values for each plasma particle).

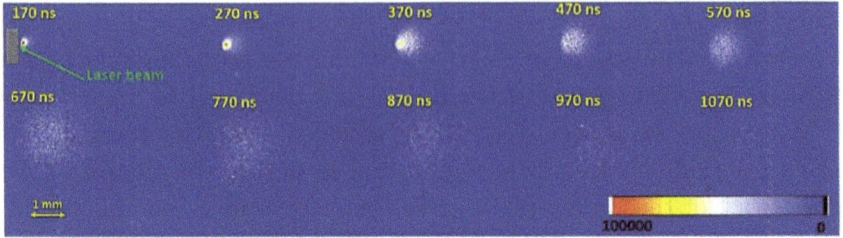

Figure 5.10. Temporal evolution of a stainless-steel LPP.

The global dynamics of the nanosecond LPP on a complex metallic target (stainless steel) has been investigated by means of ICCD fast camera imaging. The dynamic of the LPPs was investigated over a 1 μs time span (Figure 5.10). It was observed experimentally that for time delays longer than 1 μs, the emission decreases significantly, and thus, the collected data cannot be trusted. The importance of such an experimental approach is given by the opportunity to investigate possible plasma structuring in every possible direction. A prime observation is the separation of the plasma into two different structures (in the axial direction) expanding with different velocities. This splitting was previously observed for pure targets in relatively high pressures (Harilal *et al.*, 2014), and even for low background pressures for pure (Puretzky *et al.*, 2000; Irimiciuc, 2014) and complex targets (Irimiciuc *et al.*, 2017b). The separation into two or multiple plasma structures is often seen as an effect of the background gas (Geohegan, 1992) or as being induced by the different ablation mechanisms involved at various stages of ablation (Focsa *et al.*, 2017; Bulgakova *et al.*, 2004; Irimiciuc *et al.*, 2017b). However, in the past few years, there have been some reports of the heterogeneous distribution of particles in complex plasmas (Canulescu *et al.*, 2009a, 2017). They concluded that particles have different expansion velocities and can be found in particular areas of the plasma plume with respect to their atomic masses, bonding energies or thermal properties.

For our LPP, in the specified conditions of laser fluence and background pressure (50 J/cm^2 and 10^{-3} Torr), we observe a strong confined emission for the first 200 ns, described by a small volume

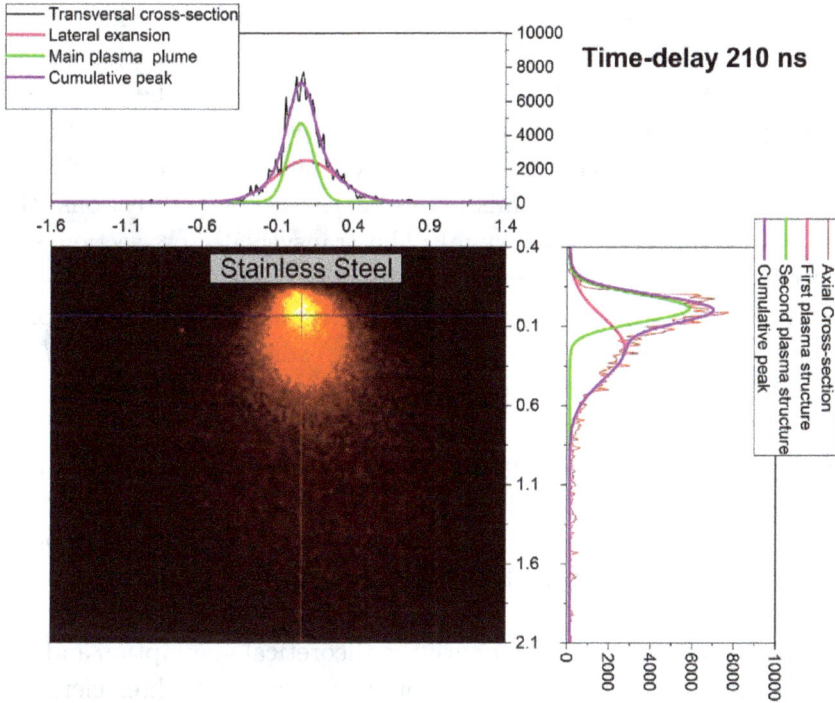

Figure 5.11. Snapshot of a stainless-steel plasma recorded after 150 ns and two cross-sections on the main expansion axis (lateral) and on the transvers axis (above).

and a high global optical emission, followed by an increase in the plasma volume, also induced by its separation into two structures (200–400 ns). After a long evolution time, the plume has a quasi-spherical shape and is defined by a low emission intensity. These characteristics were observed until the extinction of the plume after several microseconds.

Figure 5.11 presents the ICCD image of plasma collected after 210 ns and two different cross-sections performed in the axial direction and on the transverse axis. The two graphs showcase the plume-splitting process in the main expansion direction (left side). The two structures evolve with different velocities (the first plasma structure at 22 km/s and the second one at 16 km/s), a result which is in good agreement with similar studies performed using this technique (Li *et al.*, 2013; Geohegan, 1992; Irimiciuc *et al.*, 2017b). However, we note here that the velocity of the second structure is relatively higher than the ones found in the previously cited papers. The cross-section in the transversal direction reveals a symmetrical emission around the main expansion axis. We emphasize here that there is strong emission for large expansion directions (the lateral expansion of the plume). The transversal cross-section was deconvoluted with two Gauss-type functions, each representing the main plasma plume (green curve) and the lateral expansion (pink curve). This behavior, according to the work of Canulescu *et al.* (2009a), is given by the emission of low-mass elements. Thus, the same reasoning can be applied for our plasma, as the overall emission should contain several light elements, such as Mg, Ni and Cr, and heavier ones, such as Mo. This particular result was predicted by the theoretical model (see Figures 5.7(a) and 5.8(b)), where we showed that for the flow of a complex fluid, the separation occurs in its inner energy (expressed through the fractal velocity field).

The results are consistent with the theoretical assumption and the initio paradigm presented in the previous section. Lighter elements will define fractal trajectories described by a higher fractalization degree due to the increased number of collisions, a result reflected in the increase in the lateral global emission of the plume. On the other hand, heavier particles (high-mass atoms or ions, clusters and nanoparticles) will be defined by smaller fractalization degrees. Thus, an LPP generated on a complex target can be translated into a multifractal fluid. The differences in the fractality degree will define a more complex expansion of the plume in its wider regions, as represented in Figures 5.7–5.9. This separation in a velocity/energetic "space" should be reflected in the individual kinetic and thermal energies of the ejected particles. This aspect will be discussed in the following section, where space- and time-resolved OES was implemented for the study of the stainless-steel laser ablation plasma.

5.3.2.2. *Local investigations: Optical emission spectroscopy*

The OES technique was employed to differentiate between the contributions of each element to the overall shape, structure and

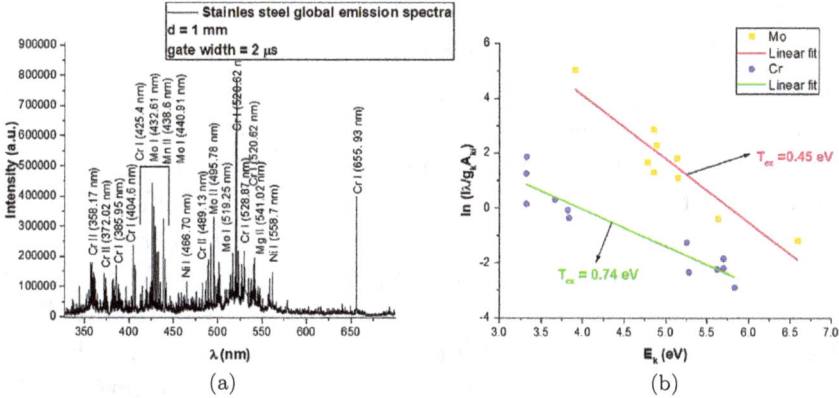

Figure 5.12. Optical emission spectra of a nanosecond LPP on a stainless-steel target (a) and the Boltzmann plot used for the temperature calculation (b).

dynamics of the stainless-steel plasma. It is based on the analysis of a 0.238 mm wide, centered plasma slice defined by the used optical setup. The optical emission spectra of the plasma plume recorded over a 2 μs gate time presented in Figure 5.12(a) allowed us to identify the emitting species by consulting a specialized database (Kramida et al., 2014). For our plasma plume, emissions from both ionic and atomic species for the same element were observed. These were associated with Cr, Ni and Mg. We notice that all these species have a similar atomic mass and are represented by a higher fractalization degree in our theoretical approach. On the other hand, for Mo, we observe only neutral lines. The absence of Mo ions from the spectra does not mean there are no Mo ions in the plume, but rather that there is a sign (among other causes) of a lower thermal energy of the electrons, which cannot excite those particular species. The (thermal) energy of the electrons is given by the excitation temperature. In a previous paper (Irimiciuc et al., 2017), we correlated the electron temperature, there determined through electrical methods, with the fractal potential of the fluid, as both parameters defined the same physical interactions. Here, the difference in the thermal energy of the individual elements is also a reflection of the difference in the fractalization degrees of the respective species.

The excitation temperatures of the individual species can be determined by using the Boltzmann plot method (Aguilera and Aragón, 2007). In Figure 5.12(b), we represent the Boltzmann plot for Cr and Mo atomic lines. Both representations could be fitted with a linear function, which can also be an indicator of an LTE. However, we observe that different species are described by different slopes and thus by different global excitation temperatures. For Mo atoms, we found the lowest excitation temperature of 0.36 eV, while for lighter elements, we found increasing excitation temperatures (Mg: 1.42 eV, Ni: 0.45 eV and Cr: 0.78 eV). This translates into thermal velocities of 1.44 km/s for Mo, 2.63 km/s for Cr, 1.46 km/s for Ni and 1.67 km/s for Mg. The results offer us an ample view of the energy distribution per each element, and we observed that lighter elements have higher thermal movements and thus higher excitation temperatures, while heavier elements are defined by low excitation temperatures and correspondingly lower thermal velocities. The experimental data sustain the proposed relationship between the plasma electron/excitation temperature and the fractalization degree of the system previously reported by Irimiciuc *et al.*, (2017). Here, we observed that the correlation is also viable for complex plasmas, assimilated with a multifractal fluid. The thermal movement of the plasma is well defined by the fractal potential and the fractalization degree. Thus, from a fractal perspective, the elements with different fractalization degrees define regions of the plume where they can be found in majority. Thus, the significant lateral emission of the plasma, seen in Figures 5.7–5.9 and 5.12, is more likely due to the presence of light elements, such as Ni, Cr or Mg (as reported by Canulescu *et al.* (2009a)), while the central part of the plume contains mainly Mo atoms.

The differences in the thermal velocities of the ejected particles coupled with the transversal separation of the plasma plume seen through ICCD fast camera imaging are in line with the results of the theoretical model. The overall flow of a complex multicomponent fluid (Figures 5.7 and 5.8) in the transversal plane with respect to the expansion direction separates into multiple regions. Each of these regions is characterized by a different fractalization degree, which corresponds to a different thermal velocity and thus a different excitation temperature. These energetic differences can affect the quality of the deposition (e.g. chemical composition, uniformity, thickness, structure). Information on these processes involved in thin-film growth becomes one of the main advantages of this approach.

5.4. Particularities of the PLD of ternary alloys

Laser ablation is a research area of increasing interest due to its wide range of applications, which include high harmonic generation (Ganeev *et al.*, 2012), micro-machining (Molian *et al.*, 2009), laser ablation propulsion (LAP) (Phipps, 2007), material synthesis (nanoparticles (Puretzky *et al.*, 2000) and thin films (Eason, 2007) and analysis. Laser-induced breakdown spectroscopy (Cremers and Radziemski, 2006) can provide an elemental analysis of samples which are difficult to study due to their locations or potential damaging effects (extraterrestrial and deep-sea samples, explosive residues). Laser ablation plasma thrusters have seen a rapid evolution due to their simplicity and small dimensions and also due to the fact that they can ensure precise controllability of the thrust. LAP remains the major electric propulsion concept, but further fundamental studies must be carried out in order to improve thrust performance (enhanced launch flexibility, on-orbit adaptability and configuration, reduced costs and weight constraints).

One of the key applications of laser ablation is the PLD technique, a powerful and versatile method used for the growth of complex nanostructured systems (nano-columnar and multilayered structures). Using this technique, almost any type of material can be deposited, and considering the numerous experimental parameters that can be changed, it offers the possibility of obtaining films with different structural, chemical, optic and magnetic properties suitable for various types of applications. In PLD conditions, the generated plasma plume is transferred to the substrate, and thus, the plasma characteristics are essential when understanding the processes implicated in the deposition. The characteristics of the laser-induced plasma can vary significantly in space and time, and thus, its diagnosis can be a difficult task. OES has proved to be a suitable, nonintrusive type of analysis, which can offer space- and time-resolved information on the evolution of the generated plasma plume as a whole and of the individual species. Despite the considerable advances made in thin-film deposition through PLD, there are still aspects related to the plume formation and expansion that are not completely understood, and many research groups are attempting to find the answers through experimental and theoretical approaches.

This study is part of an ongoing research topic of our group, which is focused on the influence of the laser characteristics (wavelength, pulse duration, repetition rate, fluence), target material, background pressure and gas type, target–substrate distance on the plasma properties and its dynamics. The reported study is based on the analysis of LPP of simple (Al) and complex targets (Ag–Ni–Fe) and how the presence of multiple types of species affects their characteristics. After observing correlations between the plasma parameters (excitation temperature, species velocity) and the single-metal properties, we extended our study to more complex materials, such as the Ag–Ni–Fe alloy. The interest in aluminum- and silver-based materials was justified by their use in photovoltaic devices as contact electrodes (Kim and Lee, 2017), nonvolatile magnetic memory cells (Kaidatzis *et al.*, 2016), light emitting diodes (Hofmann *et al.*, 2011), gas sensors (Xu *et al.*, 2000) and optic coatings, such as telescope mirrors (Fryauf *et al.*, 2016). The experimental investigations were described by a theoretical model in a fractal representation that was used to further understand the influence of target complexity on the LPP parameters.

5.4.1. *Fast camera photography*

In order to study the global evolution of the transient LPPs, the fast gated camera imaging technique was employed. The aim of these measurements was to analyze the structure and determine the expansion velocity of the plasma produced as a result of the laser–target interactions and how the target properties affect its global parameters. Figure 5.13 presents sequential images of the Al and Ag–Ni–Fe alloy LLPs in a 350 ns temporal range. As the plasmas expand, their volumes increase and the position of the center of mass (defined as the plume area with the highest emission intensity) moves toward longer distances. At short time delays, the Al plasma has a smaller volume compared to the Ag–Ni–Fe LPP, while, as it evolves, its volume increases significantly, and a splitting of the plume in the orthogonal direction is observed. A slower increase in the Ag–Ni–Fe LPP volume was observed as the plume expands.

Due to the fact that the experiments were performed at a relatively low background pressure (10^{-2} Torr), a direct relationship was observed between the displacement of the maximum emitted light area and the time delay at which the images were recoded. By plotting the two parameters and fitting them with a linear increase, the "center of mass" velocity of the plasmas was determined. When higher background pressures are used, the LPP dynamics change significantly: The plume is slowed down through collisions with the ambient gas, and the expansion of the plume is defined by a "drag"−type function (Aké *et al.*, 2006) rather than a linear one.

Figure 5.14 presents images of the Al and Ag–Ni–Fe plasmas recorded at 270 ns after the beam hits the target together with two cross-sections: one in the main expansion direction, where the axial plume-splitting phenomenon can be observed; and another one on an axis parallel to the target surface in order to observe the cylindrical symmetry of the expanding plume. The data reveal the splitting of the plume into two plasma structures with different velocities: a fast one following the front of the plume and a slow one in the vicinity of the target. The plume splitting was previously observed by our group and also by other researchers on plasmas generated on pure targets, such as Cu (Irimiciuc *et al.*, 2014), Al (Focsa *et al.*, 2017) or graphite (Harilal *et al.*, 2003), or on more complex targets (Irimiciuc *et al.*, 2017a). The formation of the two plasma structures can be explained by the existence of different ablation mechanisms. The fast structures are considered to be generated by Coulomb explosions and the slow ones by thermal mechanisms, such as phase explosions or evaporation. This result is in line with other reported data in a

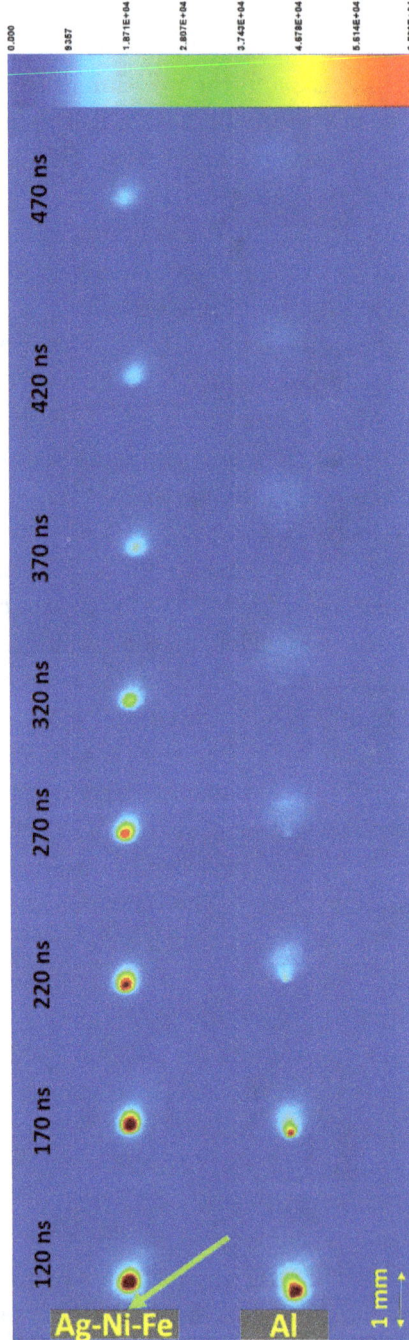

Figure 5.13. Sequential snapshots of Al and Ag–Ni–Fe alloy plasmas, recorded with a 10 ns gate width.

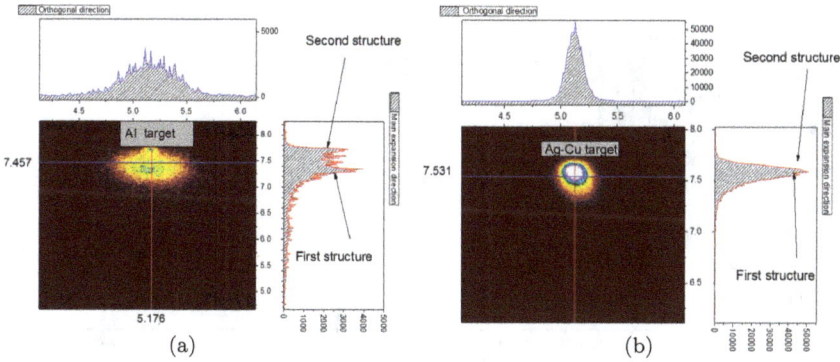

Figure 5.14. Snapshots of LPPs on Al (a) and Ag–Ni–Fe alloy (b): targets recorded with a 270 ns delay together with cross-sections in the main expansion directions and parallel to the target surface directions (inset).

significant number of papers, where the fast structures are defined by velocities of 10^4 m/s and the slow ones by 10^3 m/s. In this study, for the Al plasma, we found a velocity of 10 km/s for the slow structure and 26 km/s for the fast one, while for Ag–Ni–Fe, we found significantly higher values. For the fast structure of the plasma generated on the alloy, we recorded a velocity of 55 km/s and for the slow one, 19 km/s. These results are in good agreement with the data reported by Lippert's group (Canulescu *et al.*, 2009b; Ojeda *et al.*, 2017), where they observed two expansion velocities depending on the mass of the plume components, and are also confirmed by our previous results, where we investigated a wide range of pure metallic targets (Irimiciuc *et al.*, 2017). However, we would like to note here that the dynamics of the metallic atoms and ions ejected from pure targets are different from those of LPPs generated on complex alloys (as is the case here) due to the supplementary bonds between the alloy components (i.e. Ni–Ag), and thus, slight divergences are expected.

The main disadvantage of the ICCD imaging analysis performed so far is that it offers global information about the plume, and although the splitting process is noticeable here for the Al plasma, for the Ag–Ni–Fe alloy, one cannot observe a clear separation. Thus, we would need more insight into the evolution of the plume, and

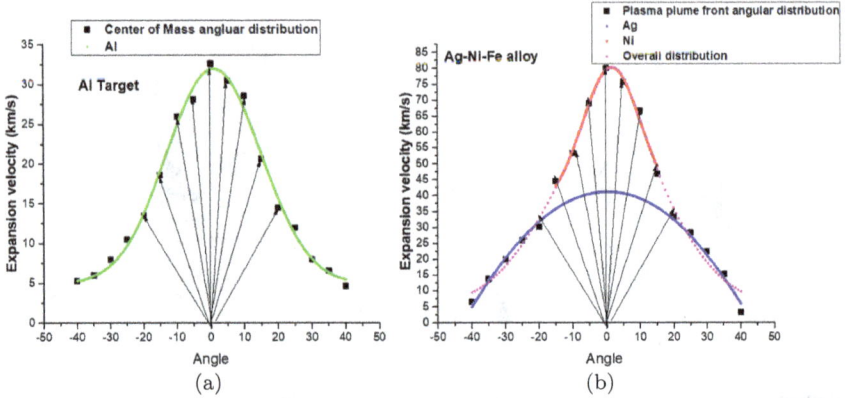

Figure 5.15. Angular distribution of the plasma plume front velocity for the Al (a) and Ag–Ni–Fe (b) investigated plasmas.

this drawback can be partially overcome by attempting to estimate the plume front velocity together with its angular dependence. The front velocity angular distributions for the two investigated plasmas are presented in Figure 5.15. The front of the plumes expands with velocities of tens of kilometers per second, having the same order of magnitude as the fast structure of each plasma. However, we notice that the absolute values of the velocities are considerable higher than the ones defined as the "center of mass" velocities. This result is somehow expected as, at every moment of time, there is a spatial distribution of the instantaneous velocity within the plasma volume, with the fastest particle defining the front and the rest defining an intense area in the vicinity of the target. The angular distribution of Al follows a function similar to the one described by the model given by Anisimov (Anisimov and Luk'yanchuk, 2002) or reported experimentally by Donnelly *et al.* (2010) and Williams *et al.* (2008). For the Al plasma, the highest velocity values are ∼30–35 km/s along the main expansion axis (0°), while for the wider regions of the plasma (40°–45°), the velocity drops down to 5–6 km/s. This strong gradient of angular velocity can explain the fast and clear separation of the structures in the Al LPP (∼200 ns), as opposed to the plasma generated on the alloy. A more complex angular distribution was noticed for the Ag–Ni–Fe LPP. Along the main expansion direction, we observe a well-defined distribution that can be described by a Gauss function; however, for larger angles, we see two symmetric humps, indicating a wider distribution of the plasma fitted by a wider Gauss function. This phenomenon can be determined by a preferential evaporation of the target component or a preferential scattering of the low-mass elements during plume expansion (Schou *et al.*, 2018; Canulescu *et al.*, 2017).

If we consider that the alloy LPP has a complex inner structure, with the lighter elements having a higher probability toward the margins of the plume (Canulescu *et al.*, 2009b), we can differentiate in our angular distribution the contribution of each element. Thus, the silver alloy plume will have a distribution focused around the main expansion direction, with about 20° reach, while the Ni one will have a wider distribution defined by lower velocities. This is due to the preferential scattering of Ni, which has a significantly lower mass with respect to Ag. Nevertheless, ICCD fast camera imaging deals with the global emission, and in order to corroborate our supposition made here, we have to investigate the dynamic of the individual elements and also look at the final product of the deposition and the quality and structure of the thin film.

5.4.2. *Optical emission spectroscopy*

The global emission analysis through ICCD imaging offers preliminary and insufficient information on the LPPs and cannot showcase the contribution of each species presented in the plume. This was done using the space- and time-resolved OES technique. For an overview of all the species present in the LPPs, global spectra were recorded with a large integration time ($\sim$2 μ s), which is plotted in Figures 5.16(a) and (b). Using the NIST database (Kramida *et al.*, 2014), the nature of the species present in the plasma was identified. The presence of both atoms and ions (including double-ionized ions) was detected. The emission intensities of the two investigated plasmas, and probably their densities, differ. This is an expected result since there is a strong difference in the energy of excited levels

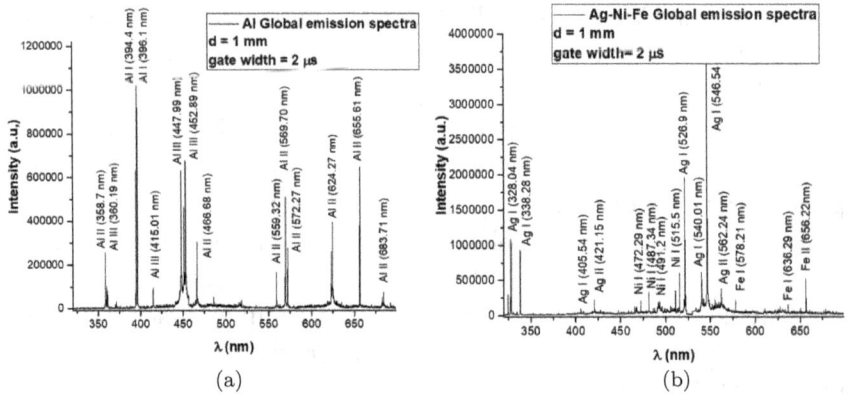

Figure 5.16. Global emission spectra recorded with a gate width of 2 μs and a gate delay of 25 ns of the Al (a) and Ag–Ni–Fe (b) LPP.

characterizing the emission lines coupled with the differences in the binding energies of Al–Al (264.3 kJ/mol), Ag–Ag (162.9 kJ/mol) and Ni–Ni (204 kJ/mol). This type of experiment suits our investigations as it showcases the effect of Ni addition (~30%) to the Ag target on the excitation process within the plasma plume.

The excitation temperatures (T_e) of all species found in the plasma were obtained using the Boltzmann plot method (Cremers and Radziemski, 2006) using the equation

$$\ln\left(\frac{\lambda I_{ki}}{A_{ki}g_k}\right) = \left[\frac{4\pi}{N_0 hc}\ln Z(T)\right] - \frac{E_k}{KT},$$

where N_0 is the total number density of atoms in the ground state, g_k is the statistical weight of the upper level, E_k is the energy of the upper level and $Z(T)$ is the partition function.

Using the data depicted in Figure 5.16 and the steps described by Aguilera and Aragón (2007), we determined the excitation temperatures of the Al, Ni and Ag atoms as ~2 eV, ~0.98 eV and ~0.2 eV, respectively. The characteristic excitation temperature of Fe was not determined since not enough atomic or ionic lines were observed. The obtained T_e values are inversely proportional to the atomic masses of the elements, which results in good agreement with the ones reported

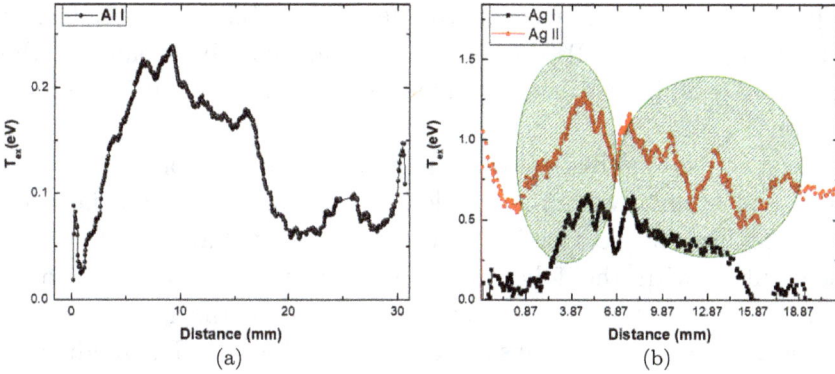

Figure 5.17. Spatial evolution of the excitation temperatures corresponding to the Al atoms (a) and the Ag atoms and ions (b) at a 30 ns gate delay.

by our group (Irimiciuc *et al.*, 2017). However, these global values cannot describe the temporal behavior of the plasma. Thus, in order to obtain a spatiotemporal mapping of the excitation temperature, we recorded the spatial distribution of specific emission lines (Al and Ag atoms and ions) at various moments in time. The results are presented in Figures 5.17(a) and (b). For both atoms and ions, we observed a steep increase in the temperature for short distances, followed by a decrease. One interesting aspect is that the spatial distribution follows a similar trend as the global emission depicted in the ICCD images. This type of dependence was observed by other groups (Wang, *et al.*, 2002; Amoruso *et al.*, 2002). A particular aspect is represented by the differences between the values of the excitation temperature of the atoms and ions of the same species, which are attributed to differential heating effect of the incoming laser pulse. The spatial distribution of the excitation temperature reveals the presence of strong fluctuations. These features can be explained in the framework of a strong modulated behavior present in the cases of high-fluence laser ablation, as it is in our case. The presence of multiple structures in LPP is generally accepted as being based on the formation of an ambipolar electrical field (single or multiple double layers) during expansion. This corresponds to the

Coulomb scenario. According to the work published by Bulgakov and Bulgakova (1999), during the first moments of expansion, the electrons will diffuse faster than the heavier particles, such as atoms or ions.

For the spatial distribution of the excitation temperature of Ag ions and atoms, we can see a clear separation at ∼7 mm between the maxima in the vicinity of the target (attributed to the slow structure), with the following one describing the fast structure. We observe that the regime overlaps well with the fast structure, while the slow structure has a *smoother* distribution. The modulated behavior can be generated by the Coulomb explosion mechanism, and thus, it will affect mainly the charged particles from plasma (found predominantly in the fast structure).

To get an idea on the location of each species in the plasma plume, we investigated the space–time evolution of the emission line characteristics of atoms and ions from the LPPs. The results are presented in Figures 5.18(a)–(d), where we plotted the signals for atoms (Al I: 396.15 nm, Ag: 546 nm and Ni: 515.5 nm) and ions (Al III: 447.5 nm). Figure 5.18(a) presents the spatial traces of both Al ions and atoms acquired after a 30 ns delay. As OES measurements were done at different distances from the target to the substrate, we observed that the first emission line detected is that of the atoms (which is indicative of a smaller expansion velocity). This is followed by the one corresponding to the ion. Such a type of analysis allows us to investigate the dynamics of each ejected particle. At the same time, we can observe that the spatial profiles of the emission lines can be placed into two main groups: the fast group — containing ions — characterized by a short evolution time and found at a larger distance from the target and the slow group — formed by atoms — with a longer evolution time and present at a smaller distance with respect to the target. Such a differentiation can be related to the temporal evolution of the plasma plume recorded by ICCD imaging, where we identified two plasma structures (fast and slow). Considering the latter results, we concluded that the fast structure contains mainly ions, while the slow structure would correspond to the ejected atoms.

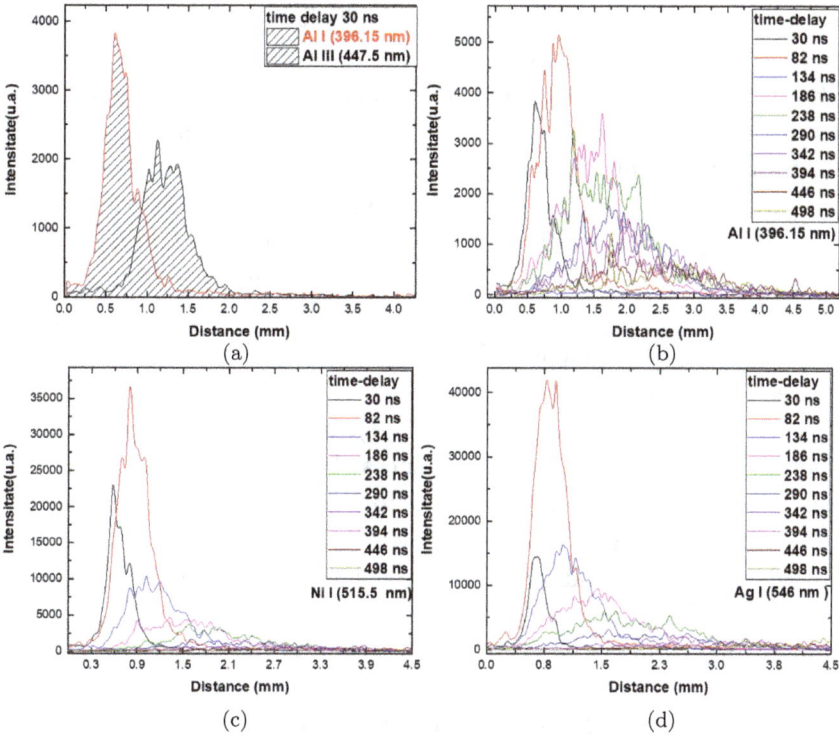

Figure 5.18. Spatial distribution of (a) Al atomic and ionic species (396.15 nm; 448 nm) and the space–time evolution of the emission lines corresponding to (b) Al (396.15 nm), (c) Ni (515.5 nm) and (d) Ag (546 nm) atoms.

In order to determine the expansion velocities of the individual species and then compare them with the ones found in the global analysis, we have represented the spatial evolution of the emission line intensity corresponding to each type of element at various moments in time. We notice that the emission intensity maximum shifts toward longer distances as the delay time is increased. By representing this spatial shift as a function of time and fitting the plots with a linear function, we determined the expansion velocity of atoms (Al: 7.7 km/s; Ni: 6 km/s; Ag: 3 km/s) and ions (Al: 16.4 km/s; Ni: 20.4 km/s; Ag: 24.7 km/s). The evolution of the atomic species velocities follow a similar trend as the one of the global velocities, and thus, it has a clear dependency on the atomic mass. This relationship between the expansion velocity and the atomic mass showcases the importance of the ejection mechanism. In general, the atomic species in LPPs are ejected through thermal mechanisms. The main thermal mechanism is phase explosion or thermal evaporation, both presenting an inverse proportionality between the velocity and the square root of mass. This is also supported by the results from Miloshevsky *et al.* (2014), where the authors reported a similar dependence but between the total charge and the atomic mass. However, when looking at the fast structure, we cannot find the same type of dependence. This is again related to the dominant ejection mechanism, which, in this case, is an electrostatic one. In a recent article (Irimiciuc *et al.*, 2017), our group showed that the conductivity of the material influences the dynamics of the charged particles. Al has a much lower conductivity (38 MS/m) than Ag (63 MS/m), which is the main element in the used alloy. Our results reveal how the fundamental ablation mechanism can influence the dynamics of the ejected particles.

5.4.3. *Thin film properties influenced by the ablation plasma dynamics*

The applications of Al- and Ag-based thin films include fields such as electronics, storage devices, renewable energy and space study instruments. The quality and properties of the film are extremely important, as they will be reflected in the overall properties of the final product. As previously mentioned, in PLD, the deposition parameters have a significant effect on the plasma characteristics and thus on the film growth, and *in situ* measurements are essential during this type of experiment. To improve the quality of the deposition, we need to understand more about the processes which take place during the plasma formation and its evolution toward the substrate surface. If the energy of the particles ejected from the target surface is low, their kinetic energy will transform into thermal momentum at the substrate surface, which will influence the thin-film growth. On the other hand, the incoming high-energy particles can enter the crystalline structure of the substrate and induce dislocations and mechanical strain, and also, a re-sputtering of the deposited particles back into the chamber can take place or even a removal of substrate particles. Besides the ejected particles which arrive at the substrate, there are other backscattered particles that redeposit on the target surface. Supplementary excitations are also induced due to their collisions with the incoming plume (Ojeda *et al.*, 2017). From the fast camera imaging, it is hard to differentiate between the possible phenomena (plume reflection and sputtering of the substrate), as we record only the nondispersed emission of the plasma plume. The dynamic of the LPP in the region of the substrate has been extensively investigated by Ojeda *et al.* (2017) in a recent paper, where they performed ICCD imaging of LPP on complex targets and observed that under high-pressure conditions, redeposition on the target and the area around the target are significant, while the deposition rate decreases. The ICCD images allowed them to observe that, for a specific target distance and background pressure conditions, the plume is reflected by the substrate, and the supplementary excitations close to the substrate area are given by the reflected particles colliding with the incoming ablated cloud. Moreover, the stopping of the plume in front of the substrate can occur if the background pressure exceeds a certain limit (Nica *et al.*, 2017; Ojeda *et al.*, 2017).

For our particular experimental conditions (high fluence, low background pressure and a small target–substrate distance), we consider that part of the presented phenomena occur during the deposition process. The SEM images of the deposited films of Al and Ag–Ni–Fe are presented in Figure 5.19(a) and (b). We notice that the Al film has a smoother and uniform distribution, while the alloy film presents a high density of micrometric clusters. The clusters in LPPs are generally induced by high fluences and are usually seen in OES measurements as black-body-type radiation, and in ICCD images, these are usually depicted as being part of another (third) structure expanding with a low kinetic velocity of about a few hundred meters per second. However, neither of the two signatures was observed in our experimental data. This leads us to the assumption that the clusters might be formed during the plume expansion. This is also supported by the interaction of the reflected plume with its thermalized tail, which leads to recombination and supplementary interactions in the vicinity of the substrate. The formation of complex molecules and nanoparticles of clusters during plume expansion has been discussed extensively by the group of Geohegan (Geohegan *et al.*, 1998; Puretzky *et al.*, 2000). They reported that in particular conditions, the plasma plume stops a few centimeters from the target, while the complex structures flow past that distance, reaching the substrate.

The chemical composition of the alloy thin film was analyzed through EDX. These measurements revealed that the film only contains a high concentration of silver, especially in the form of clusters (Figure 5.19). This segregation of the Ag particles can be explained by the difference in the cohesion energy between Ag–Ag (162.9 kJ/mol) and the bonds between other elements (i.e. Ni–Ni (204 kJ/mol)) and also by its lower heat (11.28 kJ/mol) of fusion compared to Ni (17.48 kJ/mol) and Fe (13.81 kJ/mol). Both parameters suggest that lower energies are needed for the Ag–Ag bonds to form than for those of the other elements.

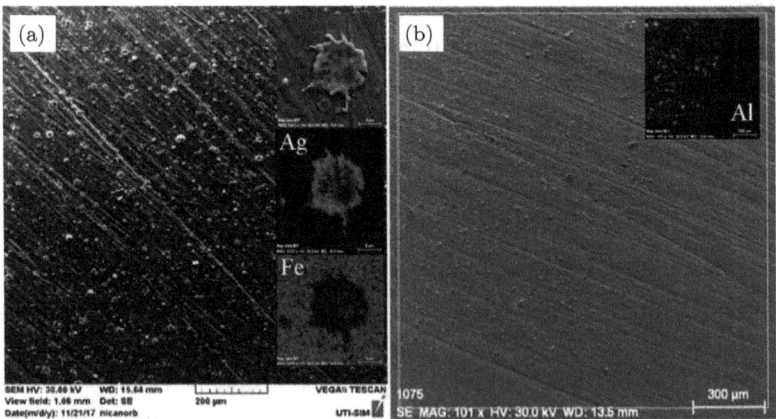

Figure 5.19. SEM images of (a) Ag–Ni–Fe and (b) Al alloy film deposited on stainless steel.

5.4.4. *A novel approach to PLD process analysis*

In order to investigate from a theoretical point of view the dynamics of plasma plumes produced on complex alloys and compare them with those generated on pure targets, we choose to work with the nondifferential model proposed by our group (Agop *et al.*, 2009) and revisited in recent years (Irimiciuc *et al.*, 2014, 2017). To achieve this aim, we use the multifractal equation (Merches and Agop, 2015; Agop and Paun, 2017). To verify the validity of such an approach, we performed numerical simulations depicting the flow of LPP with simple and multiple components. Each one of the components defines families of trajectories with a specific fractalization type (i.e. here, we considered $\lambda \sim 10$ and $\lambda \sim 1$). The first family of trajectories defines particles with higher masses and a homogeneous size distribution, having a relatively high kinetic energy. The second family defines a flow of particles with different atomic masses and a heterogeneous size distribution and properties. The first case will correspond to single-element LPPs (i.e. the case of Al, presented in the previous section), while the second one will correspond to LPPs on multi-element (complex) targets (i.e. Ag–Ni–Fe alloy).

In Figure 5.20, we have represented the total velocity field on the two main axes of the flow in a contour plot representation. For the first type of modeled plasma (Figure 5.20(a)), the dynamics are those usually assimilated to LPPs (Irimiciuc *et al.*, 2017), with a high kinetic energy that is close to the target and a quasi-exponential decrease along the expansion axis. However, for the second type of configuration, we observe that there is a segregation into multiple structures in all the expansion directions (mainly across the x-axis). There is a plasma "structure" along the main expansion axis and two symmetrically positioned secondary structures. The first confined structure is defined by a small expansion volume and a relatively high velocity, and it corresponds to elements with low fractalization degrees (Irimiciuc*et al.*, 2017), while the two lateral structures are associated with elements with high fractalization degrees. The fractalization degree is described by a geodesic shape, defined through the fluid flow of each composing element.

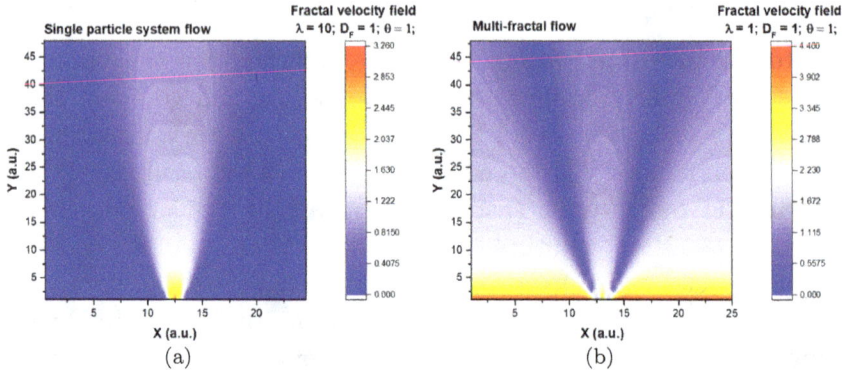

Figure 5.20. Total fractal velocity field for a single-particle system flow (a) and a multifractal system (b).

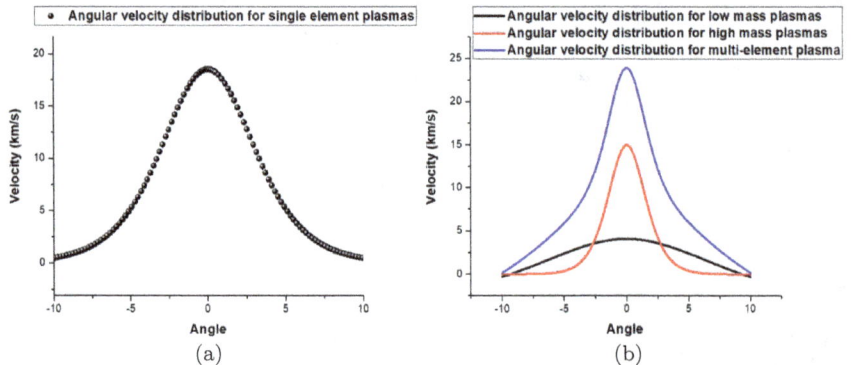

Figure 5.21. Angular distribution of the fractal velocity for simple and complex plasma plumes.

In Figure 5.21, we have represented the angular distribution of the fractal velocity of the two types of plasma translated into our fractal theoretical approach. We would like to note here that the two graphs correspond to LPPs with significantly different fractality degrees (Irimiciuc *et al.*, 2017). In Figure 5.21(a), the angular distribution for a plasma defined by a constant fractality degree (meaning a pure element in constant external conditions) is represented, and we obtain results similar to those reported by Thestrup *et al.* (2002) and Donnelly *et al.* (2010). However, when we take into account a multi-component target, we have to consider different fractality degrees for each component. Here, we emulated the experimental conditions with a bi-component target with elements having considerably different physical properties, masses and bonding energies (to best reflect the Ag–Ni configuration). We observe that the simulation performed in the framework of a fractal model gives similar results to those determined through ICCD fast camera imaging.

When attempting to combine the contributions of two elements, we had to construct a fractal fluid equivalent to the LPPs on such a target. Thus, the angular distribution needs to contain the contributions of the two elements present in the plasma plume. For the element with a lower bonding energy and a higher electrical conductivity (as is the case for Ag), the angular distribution is narrow and generally centered on the main expansion axis $(0°)$, characterized by a low fractality degree. The elements with low atomic masses, higher bonding energies and relatively lower electrical conductivities are described by a wider angular distribution characterized by a higher fractality degree.

References

Agop M., Nica P. E., Gurlui S., Focsa C., Paun V. P. and Colotin M. 2009. Implications of an extended fractal hydrodynamic model. *EPJ D*, 56(3), 405–419.

Aguilera J. A. and Aragón C. 2007. Multi-element Saha-Boltzmann and Boltzmann plots in laser-induced plasmas. *Spectrochim. Acta — Part B At. Spectros.*, 62(4), 378–385.

Aké C. Sánchez, Sanginés De Castro R., Sobral H., and Villagrán-Muniz M. 2006. Plume dynamics of cross-beam pulsed-laser ablation of graphite. *J. Appl. Phys.*, 100(5).

Amoruso S., Bruzzese R., Spinelli N., Velotta R., Wang X., and Ferdeghini C. 2002. Optical emission investigation of laser-produced MgB2 plume expanding in an Ar buffer gas. *Appl. Phys. Lett.*, 80(23), 4315.

Amoruso S, Wang X., Altucci C., De Lisio C., and Armenante M. 2002. Double-peak distribution of electron and ion emission profile during femtosecond laser ablation of metals. *Appl. Surf. Sci.*, 186, 358–363.

Anisimov S. I. and Luk'yanchuk B. S. 2002. Selected problems of laser ablation theory. Phys.-Usp., 45(3), 293–324.

Anoop K. K., Polek M. P., Bruzzese R., Amoruso S., and Harilal S. S. 2015. Multidiagnostic analysis of ion dynamics in ultrafast laser ablation of metals over a large fluence range. *J. Appl. Phys.*, 117(8).

Ausanio G., Amoruso S., Barone A.C., Bruzzese R., Iannotti V., Lanotte L., and Vitiello M. 2006. Production of nanoparticles of different materials by means of ultrashort laser pulses. *Appl. Surf. Sci.*, 252(13), 4678–4684.

Bacaita E. S., Agop M., Tarasov V. E., Zuani S., Reindl T., Rommel M., and Abdellatif M. H. 2016. A multiscale mechanism of drug release from polymeric matrices: Confirmation through a nonlinear theoretical model. *Phys. Chem. Chem. Phys.*, 18(31), 21809–21816.

Batchelor G. K. 1999. *An Introduction to Fluid Dynamics.* Cambridge University Press.

Borowitz J. L., Eliezer S, Gazit Y., Givon M., Jackel S., Ludmirsky A., Salzmann D., Yarkoni E., Zigler A., and Arad B. 1987. Temporally resolved target potential measurements in laser-target interactions. *J. Phys. D: Appl. Phys.*, 20(2), 210–214.

Bulgakov A. V. and Bulgakova N. M. 1999. Dynamics of laser-induced plume expansion into an ambient gas during film deposition. *J. Phys. D: Appl. Phys.*, 28(8), 1710–1718.

Bulgakova N. M., Stoian R., Rosenfeld A., Hertel I. V., and Campbell E. E. B. 2004. Electronic transport and consequences for material removal in ultrafast pulsed laser ablation of materials. *Phys. Rev. B*, 69(5), 054102.

Canulescu S., Papadopoulou E. L., Anglos D., Lippert Th., Schneider C. W., and Wokaun A. 2009b. Mechanisms of the laser plume expansion during the ablation of LiMn2O4. *J. Appl. Phys.*, 105(6), 063107.

Canulescu St., Döbeli M., Yao X., Lippert T., Amoruso S., and Schou J. 2017. Nonstoichiometric transfer during laser ablation of metal alloys. *Phys. Rev. Mat.*, 1(7), 073402.

Constantinescu V. N. 1984b. *Dinamica Fluidelor Compresibile.* Vol 1. Bucuresti: Editura Academiei Romane.

Cremers D. A. and Radziemski L. J. 2006. *Handbook of Laser-Induced Breakdown Spectroscopy.* John Wiley.

Cristoforetti G., Legnaioli S., Pardini L., Palleschi V., Salvetti A., and Tognoni E. 2006. Spectroscopic and shadowgraphic analysis of laser induced plasmas in the orthogonal double pulse pre-ablation configuration. Spectrochim. *Acta Part B: At. Spectros.*, 61(3), 340–350.

Cushing, B. L., Kolesnichenko V. L., and O'Connor C. J. 2004. Recent advances in the liquid-phase syntheses of inorganic nanoparticles. *Chem. Rev.*, 104(9), 3893–3946.

Dimitriu D. G., Irimiciuc S. A., Popescu S., Agop M., Ionita C., and Schrittwieser R. W. 2015. On the interaction between two fireballs in low-temperature plasma. *Phys. Plasmas*, 22(11).

Diwakar P. K., Harilal S. S., Hassanein A., and Phillips M. C. 2014. Expansion dynamics of ultrafast laser produced plasmas in the presence of ambient argon. *J. Appl. Phys.*, 116 (13): 133301.

Dizdar T. O., Kocausta G., Gülcan E., and Gülsoy Ö. Y. 2018. A new method to produce high voltage static electric load for electrostatic separation — Triboelectric charging. *Powder Tech.*, 327, 89–95.

Donnelly T., Lunney J. G., Amoruso S., Bruzzese R., Wang X., and Ni X. 2010. Angular distributions of plume components in ultrafast laser ablation of metal targets. *Appl. Phys. A: Mat. Sci. Process.*, 100(2), 569–574.

Eason R. 2007. *Pulsed Laser Deposition of Thin Films : Applications-Led Growth of Functional Materials.* Wiley-Interscience.

Focsa C., Gurlui S., Nica P., Agop M., and Ziskind M. 2017. Plume splitting and oscillatory behavior in transient plasmas generated by high-fluence laser ablation in vacuum. *Appl. Surf. Sci.*, 434, 299–309.

Fryauf D. M., Phillips A. C., and Kobayashi N. P. 2016. Corrosion protection of silver-based telescope mirrors using evaporated anti-oxidation overlayers and aluminum oxide films by atomic layer deposition. Proceedings of the SPIE 9924, pp. 1–8.

Ganeev R. A., Witting T., Hutchison C., Frank F., Redkin P. V., Okell W. A., and Lei D. Y. 2012. Enhanced high-order-harmonic generation in a carbon ablation plume. *Phys. Rev A — At. Mol. Optical Phys.*, 85(1), 3–6.

Geldart D. 1973. Types of gas fluidization. *Powder Tech.*, 7(5), 285–292.

Geohegan D. B. 1992. Fast-iccd photography and gated photon counting measurements of blackbody emission from particulates generated in the KrF-laser ablation of BN and YBCO. *MRS Proc.*, 285 (January), 27.

Geohegan D. B., Puretzky A. A., Duscher G., and Pennycook S. J. 1998. Time-resolved imaging of gas phase nanoparticle synthesis by laser ablation. *Appl. Phys. Lett.*, 72(23), 2987–2989.

Hahn D. W. and Omenetto N. 2010. Laser-induced breakdown spectroscopy (LIBS), Part I: Review of basic diagnostics and plasma particle interactions: Still-challenging issues within the analytical plasma community. *Appl. Spectros.*, 64(12), 335–366.

Harilal S. S., Bindhu C. V., Tillack M. S., Najmabadi F., and Gaeris A. C. 2003. Internal structure and expansion dynamics of laser ablation plumes into ambient gases. *J. Appl. Phys.*, 93(5), 2380–2388.

Harilal S. S., Farid N., Freeman J. R., Diwakar P. K., LaHaye N. L., and Hassanein A. 2014. Background gas collisional effects on expanding Fs and Ns laser ablation plumes. *Appl. Phys. A: Mat. Sci. Process.*, 117(1), 319–326.

Harilal S. S., Issac R. C., Bindhu C. V., Nampoori V. P. N., and Vallabhan C. P. G. 1996. Temporal and spatial evolution of C2 in laser induced plasma from graphite target. *J. Appl. Phys.*, 80(6): 3561.

Hofmann S., Thomschke M., Lussem B., and Leo K. 2011. Top-emitting organic light-emitting diodes. *Opt. Express*, 19(November), 1143–1147.

Irimiciuc S. A., Mihaila I., and Agop M. 2014a. Experimental and theoretical aspects of a laser produced plasma. *Phys. Plasmas*, 21, 093509.

Irimiciuc S., Boidin R., Bulai G., Gurlui S., Nemec P., Nazabal V., and Focsa C. 2017a. Laser ablation of (GeSe2)100−x(Sb2Se3)X chalcogenide glasses: Influence of the target composition on the plasma plume dynamics. *Appl. Surf. Sci.*, 418, 594–600.

Irimiciuc S., Bulai G., Agop M., and Gurlui S. 2018a. Influence of laser-produced plasma parameters on the deposition process: In situ space- and time-resolved optical emission pectroscopy and fractal modeling approach. *Appl. Phys A: Mat. Sci. Process.*, 124(9), 615.

Irimiciuc S. A., Agop M., Nica P., Gurlui S., Mihəileanu D., Toma Ş., and Focşa C. 2014b. Dispersive effects in laser ablation plasmas. *Jpn. J. Appl. Phys.*, 53(11).

Irimiciuc S. A., Gurlui S., Bulai G., Nica P., Agop M., and Focsa C. 2017b. Langmuir probe investigation of transient plasmas generated by femtosecond laser ablation of several metals: Influence of the target physical properties on the plume dynamics. *Appl. Surf. Sci.*, 417, 108–118.

Irimiciuc S. A., Gurlui S., Nica P., Focsa C., and Agop M. 2017c. A compact non-differential approach for modeling laser ablation plasma dynamics. *J. Appl. Phys.*, 121(8).

Irimiciuc S. A., Bulai G., Gurlui S., and Agop M. 2018b. On the separation of particle flow during pulse laser deposition of heterogeneous materials — A multi-fractal approach. *Powder Tech.*, 339, 273–280.

Kaidatzis A., Psycharis V., and Niarchos D. 2016. Sputtered tungsten — Silver and Tungsten — Aluminium thin Fi Lms for non-volatile magnetic memories applications. *Microelectron. Eng.*, 159, 6–8.

Kelessidis V. C. and Mpandelis G. 2004. Measurements and prediction of terminal velocity of solid spheres falling through stagnant pseudoplastic liquids. *Powder Tech.*, 147(1–3), 117–125.

Kim D.-J. and Ho-nyeon L. 2017. Tandemly stacked photovoltaic organic light-emitting diodes with an Al/Ag double-layer intermediate electrode. *Mol. Cryst. Liq. Cryst.*, 645(1), 185–192.

Kramida A., Ralchenko Y., Reader J., and NIST ASD Team. 2014. 2014. *NIST Atomic Spectra Database Lines Form.* (Ver. 5.2) [Online].

Lakes, R. 2009. Viscoelastic Materials. Vol. 9780521885683. Cambridge: Cambridge University Press.

Li, X., Wenfu W., Jian W., Shenli J., and Aici Q. 2013. The influence of spot size on the expansion dynamics of nanosecond-laser-produced copper plasmas in atmosphere. *J. Appl. Phys.*, 113(24).

Lin S. C., Chen S. Y., and Cheng S. Y. 2004. Characterization and composition evolution of multiple-phase nanoscaled ceramic powders produced by laser ablation. *Powder Tech.*, 148(1), 28–31.

Love A. I. J., Giddings D., and Power H. 2014. Numerical analysis of particle flows within a double expansion. *Powder Tech.*, 266, 22–37.

Mardare D., Cornei N., Luca D., Dobromir M., Irimiciuc Ş. A., Pungă L., Pui A., and Adomniţei C. 2014. Synthesis and hydrophilic properties of Mo doped TiO2 thin films. *J. Appl. Phys.*, 115(21), 213501.

Merches I. and Agop M. 2015. *Differentiability and Fractality in Dynamics of Physical Systems.* World Scientific.

Miloshevsky A., Harilal S. S., Miloshevsky G., and Hassanein A. 2014. Dynamics of plasma expansion and shockwave formation in femtosecond laser-ablated aluminum plumes in argon gas at atmospheric pressures. *Phys. Plasmas*, 21(4), 043111.

Molian P., Pecholt B., and Gupta S. 2009. Picosecond pulsed laser ablation and micromachining of 4H-SiC wafers. *Appl. Surf. Sci.*, 255(8).

Monaghan J. J. 1992. Smoothed particle hydrodynamics. *An. Rev. Astro. Astrophys.*, 543–574.

Nedeff V., Lazar G., Agop M., Eva L., Ochiuz L., Dimitriu D., Vrajitoriu L., and Popa C. 2015a. Solid components separation from heterogeneous mixtures through turbulence control. *Powder Tech.*, 284, 170–186.

Nedeff, V., Lazar G., Agop M., Mosnegutu E., Ristea M., Ochiuz L., Eva L., and Popa C. 2015b. Non-linear behaviours in complex fluid dynamics via non-differentiability. Separation control of the solid components from heterogeneous mixtures. *Powder Tech.*, 269, 452–460.

Ngom B. D., Lafane S., Abdelli-Messaci S., Kerdja T., and Maaza M. 2016. Laser-produced Sm1−xNdxNiO3 plasma dynamic through langmuir probe and ICCD imaging combined analysis. *Appl. Phys A: Mat. Sci. Process.*, 122(1), 1–7.

Nica P.-E, Irimiciuc S. A., Agop M., Gurlui S., Ziskind M., and Focsa C. 2017. Experimental and theoretical studies on the dynamics experimental and theoretical studies on the dynamics of transient plasmas generated by laser ablation in of transient plasmas generated by laser ablation in various temporal regimes. In *Laser Ablation — From Fundamentals to Applications*, T. E. Itina (Ed.). InTech.

O'Mahony D., Lunney J., Dumont T., Canulescu S., Lippert T., and Wokaun A. 2007. Laser-produced plasma ion characteristics in laser ablation of lithium manganate. *Appl. Surf. Sci. Sci.*, 254(4), 811–815.

Ojeda-G-P A., Schneider C. W., Döbeli M., Lippert T., and Wokaun A. 2017. Plasma plume dynamics, rebound, and recoating of the ablation target in pulsed laser deposition. *J. Appl. Phys.*, 121(13), 135306.

Phipps C. R. 2007. *Laser Ablation and Its Applications*, C. Phipps (Ed.), Vol. 129. Springer Series in Optical Sciences. Boston, MA: Springer US.

Puretzky A. A., Geohegan D. B., Fan X., and Pennycook S. J. 2000. In situ imaging and spectroscopy of single-wall carbon nanotube synthesis by laser vaporization. *Appl. Phys. Lett.*, 76(2), 182–184.

Qiu, Y., Deng B., and Kim C. N. 2012. Numerical study of the flow field and separation efficiency of a divergent cyclone. *Powder Tech.*, 217, 231–237.

Santos A., Lopes Barsanelli P., Pereira F. M. V., and Pereira-Filho E. R. 2017. Calibration strategies for the direct determination of Ca, K, and Mg in commercial samples of powdered milk and solid dietary supplements using laser-induced breakdown spectroscopy (LIBS). *Food Res. Int.*, 94, 72–78.

Schou J., Gansukh M., Ettlinger R. B., Cazzaniga A., Grossberg M., Kauk-Kuusik M., and Canulescu S. 2018. Pulsed laser deposition of chalcogenide sulfides from multi- and single-component targets: The non-stoichiometric material transfer. *Appl. Phys A*, 124(1), 78.

Singh J. and Thakur S. 2007. *Laser Induced Breakdown Spectroscopy*. Elsevier Publisher.

Singh J., R. Kumar, Awasthi S., Singh V., and Rai A. K. 2017. Laser induced breakdown spectroscopy: A rapid tool for the identification and quantification of minerals in cucurbit seeds. *Food Chem.*, 221, 1778–1783.

Singh S. C., Fallon C., Hayden P., Mujawar M., Yeates P., and Costello J. T. 2014. Ion flux enhancements and fluctuations in spatially confined laser produced aluminum plasmas. *Phy. Plasmas*, 21(9).

Sloyan K. A., May-Smith T. C., Eason R. W., and Lunney J. G. 2009. The effect of relative plasma plume delay on the properties of complex oxide films grown by multi-laser, multi-target combinatorial pulsed laser deposition. *Appl. Surf. Sci.*, 255(22), 9066–9070.

Thestrup B., Toftmann B., Schou J., Doggett B., and Lunney J. G. 2002. Ion dynamics in laser ablation plumes from selected metals at 355 Nm. *Appl. Surf. Sci.*, 197–198, 175–180.

Verloop J. and Heertjes P. M. 1970. Shock waves as a criterion for the transition from homogeneous to heterogeneous fluidization. *Chem. Eng. Sci.*, 25(5), 825–832.

Vitiello M., Amoruso S., Altucci C., de Lisio C., and Wang X. 2005. The emission of atoms and nanoparticles during femtosecond laser ablation of gold. *Appl. Surf. Sci.*, 248(1–4), 163–166.

Williams G. O., O'Connor G. M., Mannion P. T., and Glynn T. J. 2008. Langmuir probe investigation of surface contamination effects on metals during femtosecond laser ablation. *Appl. Surf. Sci.*, 254, 5921–5926.

Xu J., Pan Q., and Qin J. 2000. Sensing characteristics of double layer film of ZnO. 161–163.

Zhang S, Kuwabara S., Suzuki T., Kawano Y., Morita K., and Fukuda K. 2009. Simulation of solid-fluid mixture flow using moving particle methods. *J. Com. Phys.*, 228(7), 2552–2565.

Zhou L. 2017. Two-fluid turbulence modeling of swirling gas-particle flows — A review. *Powder Tech.*, 314, 253–263.

Chapter 6

Investigations on the Laser Ablation Process on Geomaterials

6.1. Introduction

One of the new emerging applications of laser ablation is laser-induced breakdown spectroscopy (LIBS), with implementation in environmental science, space applications and the food industry (Davari *et al.*, 2016). LIBS can analyze a wide range of samples, spanning from metals, semiconductors, glasses, biological tissues, plastics, soils and plants to thin-layer paint coatings and electronic materials (Hahn and Omenetto, 2010). From the perspective of analyzing geomaterials (Xu *et al.*, 2016), this technique has its advantages: namely, fast to nil preparation of samples, which also means that no special sample preparation skills are required; potential for *in situ* analysis; small sample size requirements; the samples can be in a solid (Xu *et al.*, 2016), liquid or gas state (Xie *et al.*, 2017); and sensitive to light elements, such as H, Be, Li, B, C, N, O, Na and Mg (El Haddad *et al.*, 2014). Successful applications of the LIBS technique to the analysis of a range of metallic targets (aluminum alloys, iron-based alloys, copper-based alloys and precious alloys) and a summary of good practices were discussed in the works by Palleschi *et al.* (Tognoni *et al.*, 2007). The performance of this technique is strongly dependent on the complex processes involved in plasma formation and expansion, as the study object for LIBS is the strong emitting laser-induced plasmas. As such, differential absorption by the material and by the particle vapor, or the matrix effect (Eppler *et al.*, 1996), can strongly affect the properties of the resultant plasma (usually confined to a few millimeters from the sample and expanding at high velocities). The matrix effect issues are shown to be overcome by different sample preparation techniques (i.e. in the form of pressed pelletized powder (Sanghapi *et al.*, 2016) or, with much better results, fused glass); however, these compromise the main advantage of LIBS — reduced time consumed for sample preparation — which is lost, and the operational costs are higher due to the special equipment needed. The implementation of the LIBS technique on soils also needs to consider the influences of compression force, moisture and the total content of easily ionized elements on line intensities and electron density, and because dramatic changes occur, the recommendations were to use dry samples as much as possible and observe plasma as early as it can be done in order to minimize the matrix effects on results (Popov *et al.*, 2018).

In the case of qualitative elemental analysis and to simultaneously deal with the complexity of data, statistical methods, such as partial least squares discriminant analysis (PLSDA) (Barker and Rayens, 2003; Dyar *et al.* 2012), are used to obtain information about the unknown samples.

The practicality of the technique is often shown in the little preparation of the diagnostics apparatus itself with approaches such as calibration-free LIBS (CF-LIBS) (Davari *et al.*, 2016), which manages to produce closer values than Monte Carlo simulated annealing optimization method (MC-LIBS) (Harmon *et al.*, 2013) with respect to a standard measurement. Aside their many advantages, there are still some important drawbacks to this approach, mainly the need for accurate plasma parameters of the investigated area. Real experimental conditions have proven a deviation from ideal conditions (optically thin, in local thermodynamic equilibrium (LTE) and with spatiotemporal homogeneity). Therefore, errors are introduced by experimental aberrations and the inaccuracy of spectral data, and it was found that they are contributing more to the overall uncertainty in the quantitative results than theoretical parameters (inaccuracy in the measurements of detector spectral efficiency weighs more on the results than a typical uncertainty in the electron density value) (Tognoni *et al.*, 2007). As the LIBS shifts to nonideal plasmas, one major issue is self-absorption in thick plasmas in the CF-LIBS. This phenomenon was investigated, and correction procedures were developed to ensure the reliability of results. A recursive algorithm was created to consider the nonlinear self-absorption effects occurring in the plasma, and this extended the range of applications of the CF-LIBS method (Bulajic *et al.*, 2002).

For the CF-LIBS, the presence of LTE is imperative, as such estimations of electron density and plasma temperature have a huge impact on the overall result, especially with strong heterogeneity in the ratio of the emitted lines. A special attention needs to be given to the Boltzmann plot method implemented for each composing element, which for complex minerals and geomaterials can be an arduous job, thus the need arising for combining different types of investigations for both accurate plasma parameter determination and elemental identification. Data fusions of LIBS results with complementary methods of analysis, such as Raman and reflectance spectroscopies and X-ray fluorescence, were performed on iron ore, and a remote-sensing instrument suite was integrated into the Mars 2020 Rover (Dyar *et al.*, 2012). There are also other approaches, e.g. LIBS as a quantitative analysis for materials located relatively far away from the laser source. This requirement becomes mandatory when dealing with potentially dangerous or out-of-reach radioactive or extraterrestrial materials. Mineral discrimination can also be performed with the help of the PLSDA statistical method (Dyar *et al.*, 2012).

Another domain of actual and acute interest is the detection of explosive residues using LIBS in its standoff mode of operation. Organic and inorganic explosive residues placed up to 30 m behind transparent barriers (polymethylmethacrylate and glasses) were successfully detected without false positives with eight shots as long as the laser beam energy can go through the barrier and a part of the plasma light can be collected (Gonzáles *et al.*, 2009).

6.2. Investigations of laser-produced plasmas generated by laser ablation on rocks

6.2.1. *Samples details*

The coordinates of the places from where the samples were collected are given in the following, along with information on the potential expectations in terms of composition, structure and formation taken from the literature review. This cannot reflect or aspire to fully describe our particular samples but will provide a sense of perspective to each subsequent analysis. As most of the investigated geomaterials are found to present a wide range of minerals which could characterize an entire geographical area, it becomes important not only to know the structure of our samples but also to get information on an entire family of rocks that could possibly be found in the vicinity of our collected samples.

Sample #1: The collected sample was taken from Petra, Jordan (coordinates: 30°20′11.8″ N 35°25′59.3″ E) (Waltham, 1994). The Petra area is located in Western Jordan, in a large rift valley which extends from the Gulf of Aqaba, northward along Wadi Araba to the Dead Sea and the Jordan Valley. The sample from Petra is assigned to the Umm Ishrin Formation, part of a larger sequence called the Ram Group of the Cambrian age. The sediments are deposited on the igneous, crystalline basement rock formed during the Precambrian that is exposed on the mountains outside the rift valley. Red Cambrian sandstones are about 300 m thick (Waltham, 1994). The Umm Ishrin Formation consists of medium- to coarse-grained, well- to moderately sorted and poorly cemented quartz arenites, with minor amounts of siltstone and mudstone (Makhlouf and Abed, 1991; Migoń and Goudie, 2014). The color is generally reddish brown, with hues of red, orange, yellow and white, due to the presence of iron oxides, mainly limonite. The mineralogical composition of red sandstones consists mainly of quartz (up to 95%) and then, in small amounts, feldspar and micas (Amireh, 1991; Migoń and Goudie, 2014). It is considered to be of fluvial origin (Amireh, 1991).

Sample #2: The collected sample was from Bowen Island, British Columbia, Canada (42°46′50.9″N 0°27′22.7″ W). Bowen Island is located on the west flank of the Coast Range batholith of British Columbia at the junction of Howe Sound and the Gulf of Georgia. The rocks from Bowen Island are of igneous origin, of an extrusive or intrusive nature. The rocks mainly consist of a volcanic assemblage of great thickness, made up of flows, breccias, agglomerates and tuffaceous sediments cut off by basic porphyry dykes (Leitch, 1947). Most of the rocks from the island belong to the Bowen Island Group of Lower Jurassic age (Boyle *et al.*, 1998), consisting of mafic to intermediate volcanic rocks (basalts and andesite), formed as lavas, shallow intrusions and volcanic ash deposits, interbedded with volcanoclastic sandstone, siliceous argillite, chert and tuff, and intruded by sills and shallow-level intrusions. Because they are resistant to erosion, they can form prominent hills. The volcanic formation is deformed and has tight east-trending folds, being intruded by granodiorite and monzonite of the Middle and Late Jurassic ages (Friedman *et al.*, 1990). On the eastern part of Bowen Island, where the sample was collected, massive meta-basalt flows and sills, and andesitic feldspar porphyry intrusions are found (Friedman *et al.*, 1990). On Bowen Island, dark-green, fine-grained andesites are composed of albitized plagioclase and hornblende, variably altered to form opaque minerals, epidote and calcite. These rocks are locally interbedded with thinly laminated to massive fine-grained siliceous tuffs (Friedman *et al.*, 1990). The volcanic sequence and Jurassic structures are cut off by a north-trending quartz feldspar porphyry stock of rhyodacite composition.

Sample #3: The collected sample was taken from a reddish formation near Pico Anayet (Valero Garcés and Aguilar, 1992) at 2575 m in the Pyrenees, Spain (42°46′50.9″ N 0°27′22.7″ W). The Anayet Massif is an E–W axis mountain range located on the central segment of the Pyrenean axial zone (northern Spain), displaying an extensive sedimentary record of Stephanian–Permian deposits. This large outcrop of late-Hercynian materials is surrounded by Devonian slates, Lower Carboniferous greywackes and several types of pyroclastic rocks (deformed during the Hercynian orogeny). The Permian deposits are represented by several thousand meter thick series of continental sediments as well as volcanoclastic and volcanic sediments (Valero Garcés and Aguilar, 1992). The continental detrital deposits between the Stephanian and Lower Triassic have been described as post-Hercynian molasses (Valero Garcés and Aguilar, 1992). These deposits have been classically divided into four main detrital groups, mainly composed of arenites (sandstones), conglomerates and lutites, with three basic volcanic episodes interbedded (Barnolas *et al.*, 1996).

Sample #4: The collected sample was taken from the top of Pico Collarada at 2886 m in the Pyrenees, Spain (42°42′51.88″ N, 0°28′14.81″ W). The analyzed sample was collected from the Pyrenees Mountains area, about 17 km north of Jaca, Spain. The Pyrenean orogenic belt resulted from the collision between the Iberian and the European plates during the Late Cretaceous to Miocene times. The Pyrenees consists of an axial zone, composed of Paleozoic rocks, granitoids, and metamorphic rocks, and is bounded by the North Pyrenean and South Pyrenean zones, where Mesozoic and Cenozoic rocks, clastic and carbonate rocks, are found (Dunham, 1962). The Paleogene sequence from the southern side of the Pyrenean orogen consists of Paleocene to Eocene light-colored massive limestones, turbidites (Lower–Middle Eocene flysch) and coastal, nonmarine deposits (Roigé *et al.*, 2016) (Upper Eocene — Lower Oligocene molasses). The turbidite systems (Lutetian) are built up by a succession of sandstones and mudstones, including carbonate megaturbidites (Rodríguez *et al.*, 2014).

6.2.2. *Structural investigations*

For the identification of the individual minerals found in the four samples, a series of complementary surface techniques was considered. Optical microscopy in polarized light is a fast analyzing technique which can be used to identify the individual minerals of the rocks and offers a general idea about the heterogeneity of the surface, which is one of the main factors that has to be considered

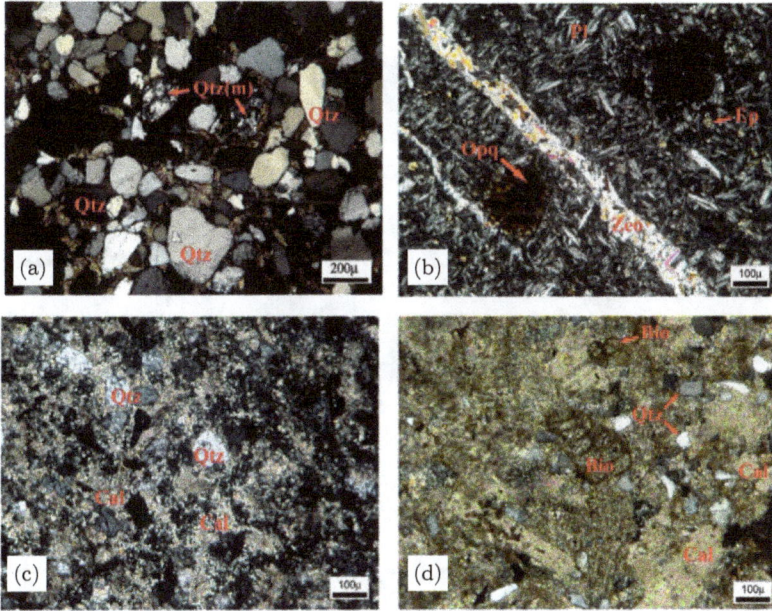

Figure 6.1. Optical microscopy and polarized microscopy images of all the investigated samples *Notes*: (Bio — bioclasts; Cal — calcite; Ep — epidote; Opq — opaque mineral; Pl — plagioclase; Qtz — quartz; Qtz(m) — metamorphic quartz; Zeo — zeolites: (a) Sample #1 — quartz sandstone, (b) Sample #2 — spilite, (c) Sample #3 — sandstone and (d) Sample #4 — extraclastic bioclastic limestone).

before a technique like LIBS is implemented. The results of these investigations are presented in Figures 6.1(a)–(d).

Sample #1 (Figure 6.1(a)) is identified as a sedimentary rock (Waltham, 1994) — quartz sandstone — that contains sand-sized grains (0.063–2 mm). The quartz grains (SiO_2) can be composed of a single crystal or can be polycrystalline (fragments of quartzite, a metamorphic rock). All grains are moderately rounded. The carbonate cement (calcite, $CaCO_3$) fills the spaces between the quartz granules. Some limonite ($FeO(OH) \cdot nH_2O$) staining can be observed, which causes the reddish color of the rock; the full picture is completed with some traces of Kaolinite ($Al_2Si_2O_5(OH)_4$) observed only through XRD (Figure 6.2(a)).

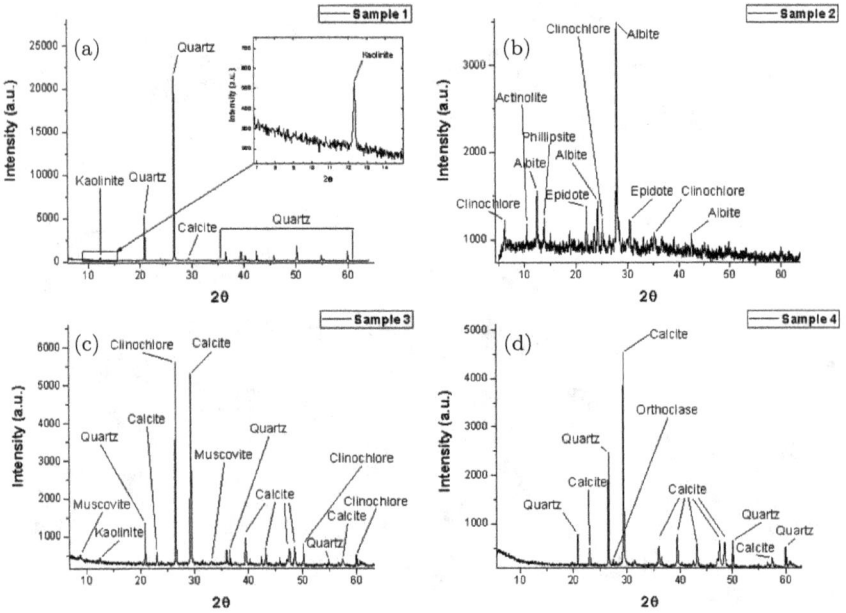

Figure 6.2. X-ray diffractogram: Sample #1 (a) — quartz sandstone; minerals: quartz, calcite, kaolinite (inset). Sample #2 (b) — Basalt (Spilite); minerals: albite (plagioclase feldspar), epidote, clinochlore, actinolite, phillipsite. Sample #3 (c) — Sedimentary rock; minerals: quartz, calcite, kaolinite, muscovite, clinochlore. Sample #4 (d) — Bioclastic limestone; minerals: quartz, calcite, orthoclase.

Sample #2 (Figure 6.1(b)) is a basalt (spilite), an extrusive igneous (volcanic) rock. Spilite is an igneous rock produced when basaltic lava reacts with seawater or is formed by hydrothermal alteration when sea water circulates through hot volcanic rocks. Under the microscope, the sample rock is composed of phenocrysts of plagioclase and very rare relict pyroxenes in a fine-grained holocrystalline groundmass made up of plagioclases, epidote $(Ca_2(Fe^{3+}, Al)_3(SiO_4)_3(OH))$, opaque minerals and iron oxides/hydroxides. Different types of alteration can be observed: primary feldspar has been transformed into albite $(NaAlSi_3O_8)$; pyroxenes and plagioclases have been replaced by other minerals, such as epidote/zoisite (in thin sections, some "nests" of epidote/zoisite surrounded by a microlitic mass consisting of the same minerals could be observed); secondary minerals as a result of alteration could be commonly chlorite ($Mg_5Al(AlSi_3O_{10})(OH)_8$) and iron oxides/hydroxides (hematite, limonite). Cavities filled with secondary minerals (probably zeolites, phillipsite: $[(K,Na,Ca)_{1-2}(Si,Al)_8O_{16} \cdot 6H_2O]$) could be observed in thin sections. Actinolite ($Ca_2(Mg, Fe)_5Si_8O_{22}(OH)_2$) (observed only on XRD, Figure 6.2(b)) is a secondary mineral as a fine-grained alteration product of pyroxene.

Sample #3 (Figure 6.1(c)) is a sandstone rock containing grains of detrital quartz (SiO_2), embedded in a matrix of carbonates, clay minerals and iron oxides/hydroxides. Due to its reddish-brown color and its composition, the sample could be a sandstone (formed in a continental environment). X-ray diffraction analysis revealed the presence of mainly quartz and calcite ($CaCO_3$), as well as small amounts of kaolinite ($Al_2(Si_2O_5)(OH)_4$), muscovite ($KAl_2(AlSi_3O_{10})(OH)_2$), clinochlore ($Mg_5Al(AlSi_3O_{10})(OH)_8$) (Figure 6.2(c)).

Sample #4 (Figure 6.1(d)) is identified as extraclastic bioclastic limestone and could be described as a bioclastic grainstone. The extraclasts are represented mainly by angular — slightly rounded quartz and orthoclase ($KAlSi_3O_8$, potassium feldspar, Figure 6.1(d)) grains, with micritic-microsparitic cement (calcite crystals are 5–10 μm in size). The fossil remnants consist mainly of foraminifera fragments (benthic and planktonic).

XRD measurements were performed on powders obtained from various areas of the samples. The obtained X-ray diffractograms are presented in Figure 6.2(a)–(d), and the identified crystalline structures are listed in Table 6.1. These results confirm the presence of the previously mentioned minerals in all samples. This confirmation allows us to have a better understanding of the interactions between the laser beam and the target, as well as how the overall ablation process is affected by the presence of such a wide spread of minerals.

The database indicatives of the minerals which fit with the XRD peaks are shown in Table 6.1.

The chemical composition and distribution of the main elements were analyzed by EDX (Table 6.2). The results revealed the presence of metallic atoms in all samples, which correspond to the elements found in the more complex minerals observed by XRD. However,

Table 6.1. Minerals confirmed by the XRD database and their indicatives.

Rocks	Confirmed minerals from XRD database
Sample #1	PDF 01-070-7244 SiO_2 Quartz
	PDF 00-058-2001 $(Al_2Si_2)_5(OH)_4$ Kaolinite 1A
	PDF 00-005-0586 $CaCO_3$ Calcite
Sample #2	PDF 00-073-2147 $Ca_2Fe_{0.33}Al_{2.67}Si_3O_{12}OH$ Ca Fe (Epidote)
	PDF 00-046-1427 $(K, Na)_2(Si, Al)_8O_{16} * 4H_2O$ (Philipsite)
	PDF 00-019-0749 $Mg_5Al(Si_3Al)O_{10}(OH)_8$ (Clinochlore)
	PDF 00-001-0739 $NaAlSi_3O_8$ (albite)
	PDF 00-080-0521 $Ca_2(Mg, Fe)_5Si_8O_{22}(OH)$ (Actinolite)
Sample #3	PDF 01-070-7344 SiO_2 Quartz
	PDF 00-024-0027 $CaCO_3$ Calcite
	PDF 01-078-2110 $Al_4(OH)_8(Si_4O_{10})$ Kaolinite
	PDF 01-070-1869 $K_{0.77}Al_{1.93}(Al_{0.5}Si_{3.5})O_{10}(OH)_2$ Muscovite-2M2
	PDF 00-007-0078 $(Mg, Fe, Al)_6(Si, Al)_4O_{10}(OH)_8$ Clinochlore
Sample #4	PDF 01-083-1762 $Ca(CO_3)$ Calcite
	PDF 01-070-7344 SiO_2 Quartz
	PDF 01-083-1324 $K_{0.59}Ba_{0.19}Na_{0.33}(Al_{0.18}Si_{2.82}O_8)$ Orthoclase

Table 6.2. Results from EDX measurements depicting each sample's elemental configuration and the respective error bar for each element.

Sample #1 (Petra)		Sample #2 (Bowen Island)		Sample #3 (Pico Anayet)		Sample #4 (Pico Collarada)	
element	at.%	element	at.%	element	at.%	element	at.%
Si	19.14 ± 0.25	Si	21.97 ± 0.25	Ca	31.85 ± 0.2	Ca	24.62 ± 0.3
Ca	5.78 ± 0.1	Fe	6.50 ± 0.13	Fe	1.33 ± 0.1	Si	17.98 ± 0.2
C	1.74 ± 0.1	Al	9.31 ± 0.16	C	0.27 ± 0.05	Al	2.84 ± 0.1
Al	2.14 ± 0.12	Mg	5.03 ± 0.11	Si	1.00 ± 0.08	Mg	0.93 ± 0.08
Ti	0.23 ± 0.05	Ba	0.68 ± 0.1	Al	0.6 ± 0.05	K	4.39 ± 0.1
Fe	0.68 ± 0.05	Ca	2.04 ± 0.11	Mn	0.67 ± 0.06	Fe	0.81 ± 0.06
Mg	0.53 ± 0.05	K	1.63 ± 0.08	K	1.47 ± 0.08	C	0.34 ± 0.05
O	69.71 ± 1.1	Na	0.73 ± 0.07	O	62.77 ± 1.2	O	48.06 ± 1.0
		Cl	0.58 ± 0.05				
		P	1.05 ± 0.07				
		C	1.13 ± 0.05				
		O	49.27 ± 1.0				

for Sample #1, traces of Ti and Mg were noticed, which can be considered impurities due to the environment and natural conditions. Similar descriptions can be made for all samples, as all of them present a more heterogeneous distribution of elements on the surface.

Two of the most important elements in the Earth's crust are silicon (Si) and calcium (Ca), which are part of the various minerals (silicates and quartz, and calcite, respectively), so we focused on the concentrations of these two elements. From Table 6.2, it can be observed that Sample #2 contains the lowest concentrations of calcium (~6%), while Sample #3 contains the lowest concentration of silicon (1%) and the highest concentration of calcium (~32%). Thus, it is safe to assume that Sample #2 may contain a very low concentration of calcite and that Sample #3 contains low concentrations of silicates but high concentration of calcite. Sample #2 contains high concentrations of silicon (~18%), iron (~6.5%) and aluminum (~9%), which points to the inclusion of significant quantities of alumino-ferro-silicates.

For all investigated samples, we performed an estimation analysis related to the microporosity of the surface. We determined microporosities between 4% and 12% across all sample surfaces (Sample #1: 12%, Sample #2: 4.7%, Sample #3: 9.23% and Sample #4: 4.9%). The porosity was determined over an average of ten surfaces of 1 mm^2, with the assumption that throughout the surface of the samples, the values of microporosity were similar. These differences in the porosity of the targets could be expected to influence the values of the expansion velocities of the laser-produced plasmas (LPPs).

6.2.3. *Optical investigations of laser-produced plasmas*

When a laser beam impinges on a surface, the beam energy is absorbed by the target. For the case of homogenous materials, the energy is transferred to the electrons, which are the first species ejected from the target by means of Coulomb explosion, while the rest of the target goes through various phase changes from solid to liquid to vapor, thus completing the ablated cloud, which represents the object of study for LIBS. As expected in the case of our samples, the beam energy will be absorbed differently by the various minerals presented in the target. These phenomena can lead to a complex ablation process, which is harder to analyze than in the case of samples of single elements or even minerals.

The general LIBS technique is used for the identification of elements from a sample. More information (in terms of plume center-of-mass velocity and wave front dynamics) can be obtained by acquiring ICCD images at various delays. This approach was implemented in the study of simple (Irimiciuc *et al.*, 2014) or complex targets (Irimiciuc *et al.*, 2017) in controlled conditions.

The fast camera imaging is suitable for transient phenomena investigations. In the case of LPPs, it is necessary to have an adequate triggering system as LPPs generally have a lifetime of a few microseconds (Puretzky *et al.*, 1993). Each recorded image is generally described by a series of parameters: resolution (which is given by the CCD detector and the optical system), time delay (the moment of time, with respect to the trigger signal, at which acquiring starts) and the gate width (or integration time). In this case, the initial moment ($t = 0$) is considered to be the "laser beam — target interaction moment". In order to have a good temporal resolution, the gate width is usually a few nanoseconds, and it can increase toward longer evolution times, where the plume is more rarefied and the emission is weaker. After their recording, the images are transferred to the computer, where they can be further analyzed. In order to estimate the expansion velocities, bi-dimensional images ("snapshots") of the LPPs were recorded at a constant laser fluence ($19 \ J/cm^2$) at various moments in time with respect to the laser beam (Figure 6.3). During the expansion, the LPP increases its volume and the center of mass, estimated as the maximum emission intensity zone, shifts toward higher distances as the recording time is changed. This leads to the conclusion that the expansion velocity is constant during the whole lifetime of the plume.

The cross-section in the expansion direction of the recorded images (Figure 6.4) shows more clearly the presence of two maxima, which were attributed to two plasma components (Amoruso *et al.*, 1999). Due to the difference in their expansion velocities, in the literature, they can be found as the fast structure (or the "first structure") and the slow structure (or the "second structure")

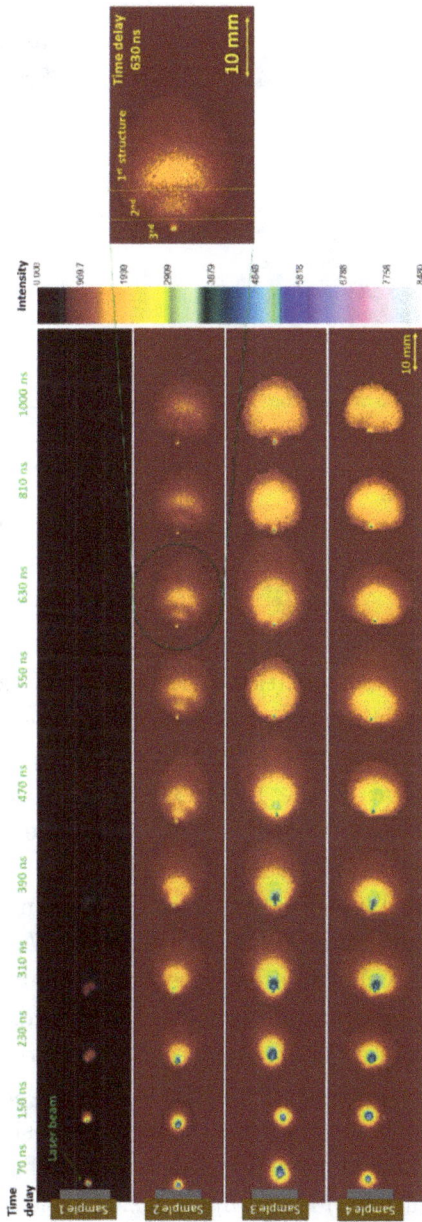

Figure 6.3. ICCD camera images, on a range of 1 μs, of LPPs generated by nanosecond laser ablation of mineral samples, and a detail of the three plasma structure (inset).

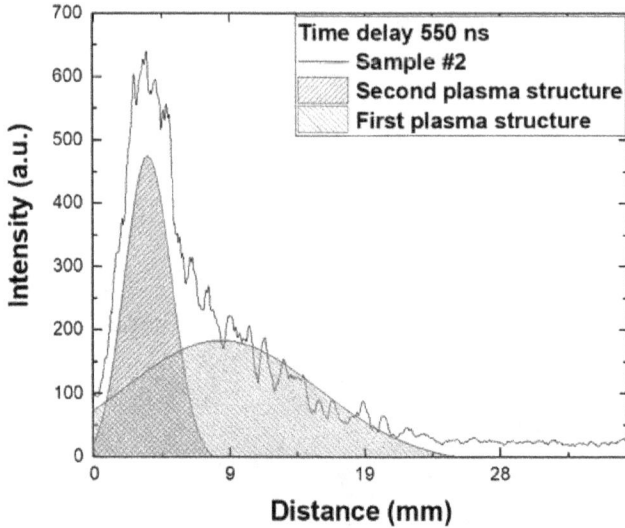

Figure 6.4. Cross-section on the ICCD snapshot collected at 550 ns of the LPP generated on Sample #2.

(Irimiciuc *et al.*, 2017). Each of the two plasma structures expands with constant but different velocities. The velocity of each structure was determined using the distance over time representation of the maximum intensity characteristic to each structure, a method in line with the theoretical view of the LPP expansion at low pressures. The constant nature of the expansion is given by the linearity with respect to all the investigated plasmas.

The plume-splitting behavior has been experimentally reported by several groups (Amoruso *et al.*, 2004; Harilal *et al.*, 2002), and it is considered a result of the different ejection mechanisms involved in the nanosecond laser ablation process. Therefore, the first, or fast, structure of the plume is ascribed to the electrostatic ejection mechanism (Coulomb explosion), while the second, or slow, structure corresponds to thermal mechanisms (phase explosion, explosive boiling, evaporation). Our results show the presence of a third plasma structure (inset in Figure 6.3) for Samples #2, #3 and #4, described by a small emission region in the proximity of the target. In the literature, this structure is attributed to the presence of clusters, nanoparticles or molecules, and it has its origin in the Knudsen layer, which is usually characterized by black body radiation (Harilal *et al.*, 2003; Chen and Bogaerts, 2005; O'Mahony *et al.*, 2007). The structures observed in this study can be compared to the structures observed in ablations on Ni, Al, stainless steel (Irimiciuc *et al.*, 2014, 2018) or more complex targets, such as GeSe chalcogenide glasses (Irimiciuc *et al.*, 2017).

For Samples #1 and #2, the expansion velocities of the first structure are of the order of tens of kilometers per second (Sample #1: 11 km/s and Sample #2: 13.5 km/s), while for the second structure, we found the velocities to be in the order of a few kilometers per second (Sample #1: 4 km/s and Sample #2: 6.5 km/s).

Although the velocity of the second structure of the plasma plume generated on Samples #3 and #4 is in line with the values of the previous samples (Sample #3: 3 km/s and Sample #4: 6 km/s), for the first structure, we found relatively low velocities (Sample #3: 8 km/s and Sample #4: 9 km/s), which is most probably related to the nature of the mineral in each sample, which could enhance the thermal ablation mechanism to the detriment of the electrostatic ones. This aspect of the LPP is consistent with other reported results on pure metals or other complex materials, and it is important for the LIBS technique, as a major part of the emission is given by this second thermalized structure. For the second structure, in the case of nanosecond laser ablation, the emission is enhanced by the absorption of the laser beam tail by the ejected particle cloud.

With respect to the expansion velocity values of the first and second plasma structures, the highest ones were found for the plasma generated on Sample #2 (Figure 6.5), the one that contains almost no calcite. The lowest velocities were observed for the plasma plumes of Sample #3, the one that contains very low concentrations of quartz. All the other plasmas (generated on the sample with both calcite and quartz) were found to be expanding with intermediate velocities. We note, however, that no correlation was observed between the porosity of the target and the estimated velocities. Samples #2 and #4 presented similar porosities but strong differences in expansion velocities for each of the two plasma structures. A similar observation

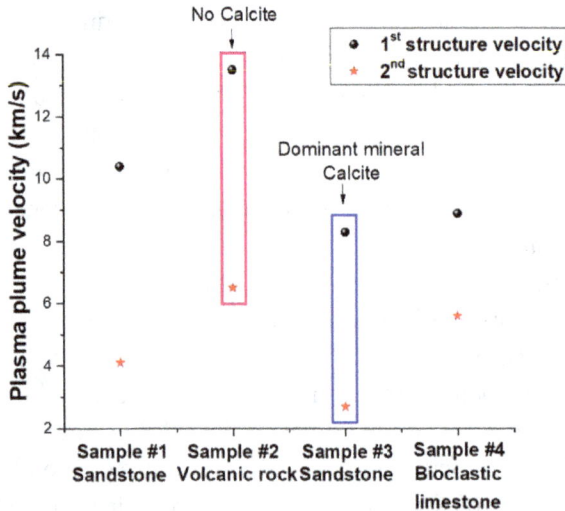

Figure 6.5. Plasma structure velocities for Samples #1, #2, #3 and #4 and their dependence on the abundance of calcite structure.

can be made for Samples #3 and #1, where, for approximately similar porosities, consistent differences in the expansion velocities were observed.

A possible explanation for this behavior can be found by analyzing the energy transfer during laser–target interactions. For samples where strong bonds, such as C=O or Ca–O, are present, more of the incident laser energy is used to break the bonds, leading to some relatively slower plasmas. This is the case for Samples #1, #3 and #4. These differences can be seen as a signature of the petrographic origins of the investigated rocks: Sample #2 is part of the magmatic rock family, while the other three samples belong to the sedimentary family — Sample #1 and Sample #3 are sandstones, while Sample #4 is a bioclastic limestone.

The optical emission spectroscopy technique (Ershov-Pavlov *et al.*, 2008) can help determine the nature of the ejected particles through the energetic levels by identifying the wavelength and by using specialized databases (Kramida and Ralchenko, 2014). The profile and intensity of the spectral lines can also provide information regarding the interactions between the ejected particles (e.g. Stark broadening (Rao *et al.*, 2016)) and the internal energy of the plasma (i.e. electron temperature and electron density). For the plasma generated on each target, the global emission spectra were collected using a gate width of 2 μs. The experimental configuration ensures the collection of a 600 μm plasma slice centered on the main expansion direction, thus providing a global characteristic in both spatial and temporal perspectives. This was done in order to collect all the emission lines, regardless of their flight time (Puretzky *et al.*, 1993). For each of the samples, we have identified atomic and ionic species characteristic of Ca, Si, Al, Mg, C and O. Samples #2 and #3 have revealed the presence of their elements, such as K, Cl or S (see Figure 6.6). Most of the emission lines correspond to the elements identified using the EDX and XRD techniques, as discussed in the previous section. Thus, a qualitative comparison can be done between

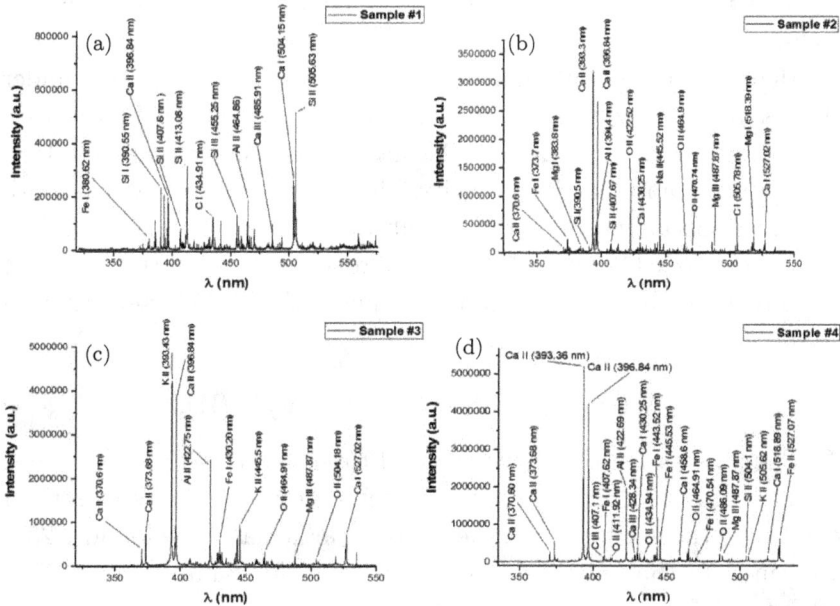

Figure 6.6. Global emission spectra collected at 1 mm from the target with a 2 μs gate width, gate delay of 100 ns and laser fluence of 19 J/cm^2 of all the investigated samples.

the LIBS signal and the EDX measurements. We notice that the evolution trends observed from EDX follow the LIBS signal changes; as such, the increase in Ca amount by a factor of 17 in the target would lead to an enhancement of the Ca line intensity by a factor of 19. For the case of Si, an increase by a factor of 26 would only lead to an increase of a factor of 6. Finally, for all the other elements, which were found in significant smaller amounts (such as Al or Fe), an eight times higher concentration would lead to an increased emission lines intensity of approximately eight times. We observe that although the ratio is not always maintained, especially for Si or lighter elements, such as K, most of the elements follow the changes from the target. At this moment, a quantitative proportionality is difficult the be achieved between the LIBs signal and the EDX data, given the complex processes involved in both investigation techniques. However, under LTE, the LIBS line intensity depends on the concentration of neutrals and singly ionized species, which allow us to tentatively use a qualitative comparison between the two techniques.

Most, if not all, plasma diagnostic techniques are valid under the assumption of the presence of an LTE. However, in the case of transient plasmas, all the plasma parameters, such as electron temperature and particle density, show a steep decrease in both time and space, and thus, the equilibrium has to be understood in a dynamic mode. There are different approaches to estimating the limit of LTE, with the most common one being the McWhirter criterion (Rao *et al.*, 2016):

$$N_e(\text{cm}^{-1}) \geq 1.6 \times 10^{12} \Delta E^3 (\text{eV}) T_e^{\frac{1}{2}} (\text{K}).$$

The above relation provides a real threshold above which we can assume LTE and implement the investigation techniques to further determine the excitation temperatures. Particularly, we found $1.23 \times 10^{15} \text{cm}^{-3}$ for Sample #1, 1.5×10^{15} cm^{-3} for Sample #2, 1.25×10^{15} cm^{-3} for Sample #3 and finally 1.18×10^{15} cm^{-3} for Sample # 4.

Once established, the limit for which LTE model can be applied, and the electron density can be estimated from the Saha–Eggert equation (Aguilera and Aragon, 2007). The relationship connects the plasma ionization equilibrium temperature to the proportion of the population of two successive ionization states:

$$n_e = 4.83 \times 10^{15} \frac{I^* g^+ A^+ \lambda^*}{I^+ g^* A^* \lambda^+} T_e^{1.5} e^{-\frac{V^+ + E^+ - E^*}{k_B T_e}},$$

where the $(*, +)$ superscripts represent the neutral excited atom and the singly charged ion, respectively, I is the emission intensities of a spectral line of λ wavelength (nm), T is the ionization temperature (expressed in K), which is taken as the excitation temperature in LTE conditions, V^+ is the first ionization potential and E is the energy of the upper level of the transition. By implementing the Saha–Eggert equation, we found a series of global n_e values for our plasmas that characterizes the 600 μm wide plasma volume throughout its evolution: 1.5×10^{16} cm^{-3} for Sample #1, 5.5×10^{16} cm^{-3} for Sample #2, 8.2×10^{16} cm^{-3} for Sample #3 and respectively 1.6×10^{17} cm^{-3} for Sample #4. The electron density exceeds the LTE limit by about one order of magnitude; thus, within our experimental conditions of laser fluence, background pressure and acquisition parameters, all the investigated plasmas respect the McWhirter criterion for LTE.

The excitation temperature of the plasma can be simply calculated using the method given by Aguilera and Aragon (2007), which uses the intensity ratio of two spectral lines characterizing the same species (ion or atom), with the specific spectroscopic data (E, A, f) found in various databases (e.g. Kramida and Ralchenko, 2014). We note, however, that there are some reservations regarding the latter parameters, which can lead to significant uncertainties. In order to minimize the errors regarding the values of the oscillator strengths, it is suitable to use not two but a series of atomic lines with different upper excitation levels. The Boltzmann plot method represents the logarithmic function of the line intensity versus the upper-level energy:

$$\ln\left(\frac{I_{ki}\lambda}{g_k A_{ki}}\right) = \ln\left(N_0 \frac{hc}{4\pi Z(T)}\right) - \frac{E_k}{k_b T_e}.$$

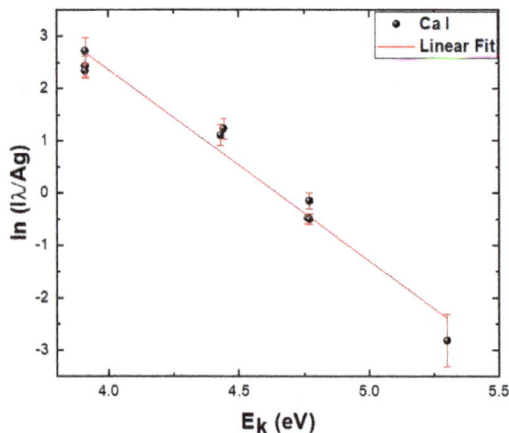

Figure 6.7. Representative Boltzmann plot for the Ca atoms representing the plasma generated on Sample #2.

The slope of this representation will give the excitation temperature, and its linearity or the deviation from it can be considered an indication of LTE validity (an example can be seen in Figure 6.7).

The values of the excitation temperature were found to be in a range of 0–1 eV for the atomic species (i.e. Ca in Sample #1: 6264 K, Sample #2: 9976 K, Sample #3: 6496 K and Sample #4: 5800 K) and of about one order of magnitude higher for the ions (i.e. Ca in Sample #1: 39440 K, Sample #2: 15080 K, Sample #3: 32480 K and Sample #4: 41760 K). These discrepancies were previously reported by our group (Irimiciuc *et al.*, 2017, 2018, 2019), where they were related to the differential heating of the plume by the incoming laser beam. However, under LTE conditions, we would expect that, regardless of the nature of the atom/ion investigated, the plasma temperature should have the same values. For all the investigated atoms, the values of the temperatures are almost the same (within a 5% error margin), while for the ions, the discrepancies are higher. For the identified metallic species (Fe, Al, Ti and Mg) in the plasmas, we found electron temperatures between 3480 and 5800 K through all the investigated plasmas, while for lighter elements, such as Ca, C, Si or K, we found significantly increased temperatures (from 6264 K for Ca up to 19720 K for K). The excitation temperatures were found to be relatively similar to the ones obtained on ablation on copper (13200–17200 K) and lead (11700–15300 K) (Lee *et al.*, 1992). For a silicon target, the study by Milan and Laserna (2001) found that the electron temperatures varied from 6000 K to 9000 K.

The plasmas generated on Samples #1, #3 and #4 (all sedimentary rocks) present the highest values for the excitation temperature for the ionic species of Ca and Si with respect to the other investigated plasmas. This is in line with the structure observed through EDX investigation, as all three targets have significantly larger concentrations of Ca. The result also correlates well with the values of the expansion velocity of the first plasma structure, determined through ICCD fast camera imaging. On the other hand, the plume generated on Sample #2 (volcanic rock) presents the highest values for the neutral species, having the highest expansion velocity for the second plasma structure. In the literature (Irimiciuc *et al.*, 2017, 2018, 2019), the first plasma structure contains mainly ions, while the second and third ones contain mainly neutral species. This view of the nature of the plasma components is also confirmed by our experimental results. The spectral investigations revealed that the temperature of the ionic Ca is much lower than that of the ionic Ca for the other samples. This could be related to the nature of Sample #2 — a magmatic rock that has almost no calcite in it versus the other samples, which are sandstones.

The experimental data showcased that the properties of the LPP strongly depend on the nature of the rocks. Nevertheless, the nature of the rocks includes multiple variables, such as the composition of minerals, the elemental composition of each mineral and the physical properties of the rock resulting from the mechanisms of mineral formation across time, which were addressed in our study by performing XRD and optical microscopy. However, the quantity of calcite microcrystals is not directly responsible for low-velocity plasma plumes, as there are other aspects to be considered, such as the porosity of the rock, the degree of impurity on the sample, crystallinity, grain size, hardness, coherence, moisture and reflectivity of the targe. The state of the sample is one major aspect that needs to be considered in order to have better control over the LIBS process.

Thus, the extension toward quantitative analysis in terms of atomic or ionic temperature and expansion velocity becomes a strong tool for material investigations and understanding how the *history* of the target affects the laser–matter interaction process and subsequently the LIBS technique.

6.2.4.   *Fractal analysis*

In the following, a fractal analysis will prove a multi-structuring of the ablation plasma in the form of Coulomb, thermal and cluster structures. Let us consider the solutions for the fractal hydrodynamic equations system in the following form Chapter 3. In this context. using the normalizations

$$\frac{x}{\alpha} = \xi, \quad \frac{V_0 t}{\alpha} = \tau, \quad \frac{V_D}{V_{D0}} = \overline{V}_D, \quad \frac{V_F}{V_{F0}} = \overline{V}_F,$$

$$\frac{\rho}{\rho_0} = \overline{\rho}, \quad \left(\frac{\lambda}{\alpha V_0}\right) = \theta, \quad V_0 = V_{D0}, \quad \frac{\lambda}{\alpha} = V_{F0}, \quad \rho_0 = \frac{1}{\alpha \sqrt{\pi}},$$

$$(6.1)$$

the differentiable velocity, the non-differentiable velocity and the states density take the forms

$$\overline{V}_D = \frac{1 + \theta^2 \xi \tau}{1 + \theta^2 \tau^2}, \tag{6.2}$$

$$\overline{V}_F = \frac{\theta(\xi - \tau)}{1 + \theta^2 \tau^2}, \tag{6.3}$$

$$\overline{\rho} = \frac{1}{(1 + \theta^2 \tau^2)^{1/2}} \exp\left[-\frac{(\xi - \tau)^2}{1 + \theta^2 \tau^2}\right]. \tag{6.4}$$

From (6.2) and (6.4), the state current density at a differentiable scale resolution takes the form

$$\overline{j}_D = \overline{\rho} \overline{V}_D = \frac{1 + \theta^2 \xi \tau}{(1 + \theta^2 \tau^2)^{3/2}} \exp\left[-\frac{(\xi - \tau)^2}{1 + \theta^2 \tau^2}\right], \tag{6.5}$$

while the state current density at a fractal scale resolution takes the form

$$\overline{j}_F = \overline{\rho} \overline{V}_F = \frac{\theta(\xi - \tau)}{(1 + \theta^2 \tau^2)^{3/2}} \exp\left[-\frac{(\xi - \tau)^2}{1 + \theta^2 \tau^2}\right]. \tag{6.6}$$

During the expansion of an LPP, we can identify three important moments (chronologically): the Coulomb explosion moment, the thermal ejection moment and the cluster formation moment. Each of these moments is defined by three different types of ejection mechanisms, which lead to the formation of three independent plasma structures (Harilal *et al.*, 2002; Chen and Bogaerts, 2005). In such a context, the dynamics of the fast plasma structure generated though the Coulomb explosion mechanism would be described by relations (6.2), (6.4) and (6.5), while the dynamics of the slow structure generated through the thermal ejection mechanisms are given by relations (6.3), (6.4) and (6.6). The reasoning behind this association is given by the fact that the non-differentiable behavior of the laser ablation plasmas is induced by the collision process between the ejected particles in each plasma structure. In Figure 6.8, we have shown the 3D representation of the states current density at differentiable and fractal resolution scales for different degrees of fractalizations, which is identified with the real particle density. With an increase in the fractalization degree, we observe a change in the slope defining the velocity of the differentiable part of the current. This can be read as an increase in the thermal velocity of the particle, as this component is induced by thermal mechanisms. On the other hand, the particle current at a fractal resolution scale, induced by Coulomb mechanisms, will present two components. These components are generated by a double layer formed in the initial stages of ablation. The fractality degree has little contribution to the spatiotemporal evolution of this component, although at a higher resolution scale, we observe a better separation between the two components of the current.

In order to describe the dynamics of the third substructure, containing mainly clusters and nanoparticles, we postulate that the specific momentum at the global scale resolution is null. This means that at the differentiable scale resolution, the velocity is equal and of opposite sign to the velocity at the fractal scale resolution (Agop and Paun, 2017):

$$V_D = -V_F = \lambda(dt)^{\left(\frac{2}{D_F}-1\right)} \partial_x(\ln \rho). \tag{6.7}$$

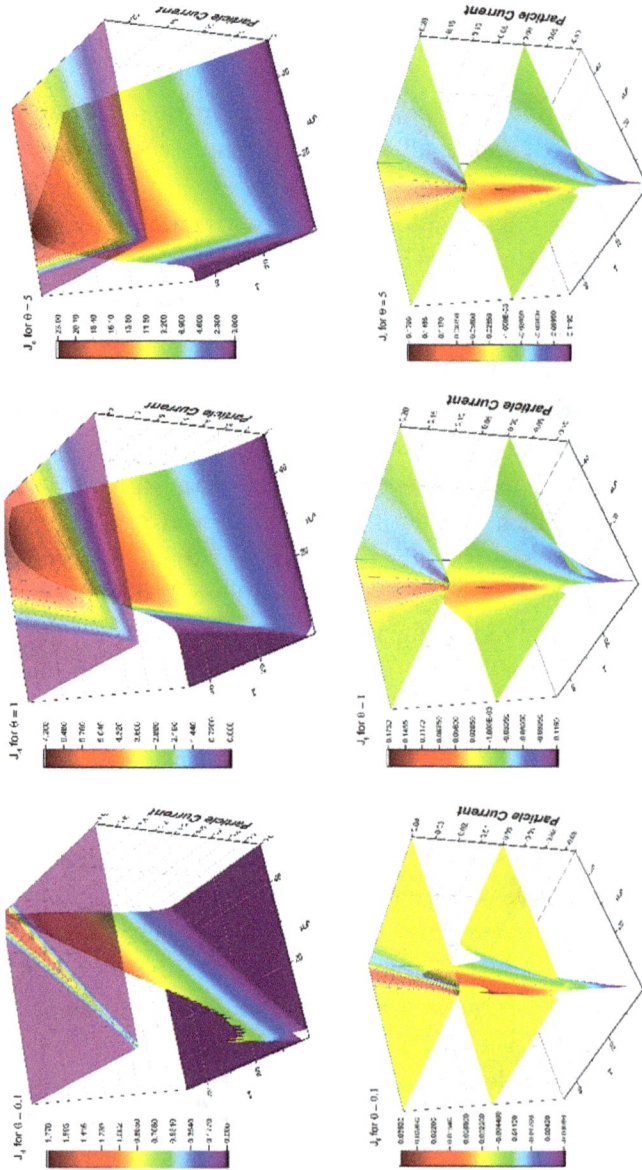

Figure 6.8. 3D representations and contour plots of the differentiable and fractal particle densities for various degrees of fractalization ($\theta = 0.1$, 1 and 5).

In these conditions, the conservation law of the state density,

$$\partial_t \rho + \partial_{xx} (\rho V_D) = 0.$$

takes the form of a fractal diffusion equation,

$$\partial_t \rho = \lambda(dt)^{\left(\frac{2}{D_F}-1\right)} \partial_{xx} \rho. \tag{6.8}$$

The solution to this equation has the following expression (Irimiciuc *et al.*, 2019):

$$\rho(x,t) = \frac{a}{\left(4\pi\lambda(dt)^{\left(\frac{2}{D_F}-1\right)}t\right)^{1/2}} \exp\left[-\frac{(x-b)^2}{4\lambda(dt)^{\left(\frac{2}{D_F}-1\right)}t}\right], \tag{6.9}$$

where a and b are integration constant. In such a context, the velocity takes the form

$$v = \frac{x-b}{2t}, \tag{6.10}$$

while the states current density is

$$j = \frac{a(x-b)}{\left(16\pi\lambda(dt)^{\left(\frac{2}{D_F}-1\right)}\right)^{1/2} t^{3/2}} \exp\left[-\frac{(x-b)^2}{4\lambda(dt)^{\left(\frac{2}{D_F}-1\right)}t}\right]. \tag{6.11}$$

Now, we calibrate the structure that contains mainly clusters with the dynamics of the other two structures in order to admit a normalization by imposing the restrictions: $a \equiv 1$ and $b \equiv 0$. We can find

$$\bar{\rho} = \frac{1}{(4\theta\tau)^{1/2}} \exp\left[-\frac{\xi^2}{4\theta\tau}\right], \tag{6.12}$$

$$\bar{v} = \frac{V_D}{V_0} = \frac{\xi}{2\tau}, \tag{6.13}$$

$$\bar{j} = \frac{\xi}{(4\theta\tau)^{1/2}\tau^{3/2}} \exp\left[-\frac{\xi^2}{4\theta\tau}\right]. \tag{6.14}$$

In Figure 6.9, we have represented a 3D representation of the current density for various degrees of fractalization depicted through θ. The three ranges of values were chosen for the fractalization degree

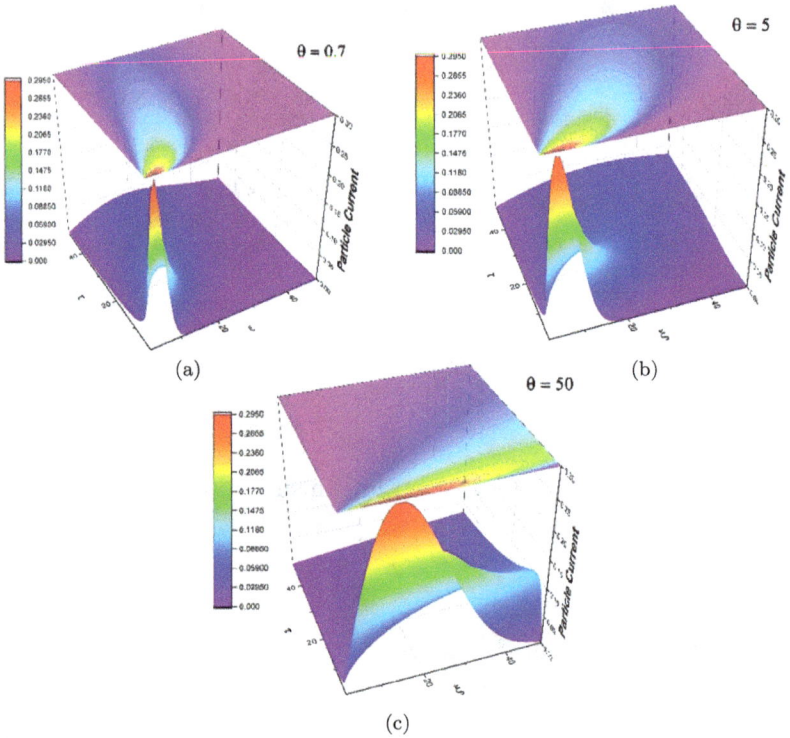

Figure 6.9. 3D representations and contour plots of the global particle density at various degrees of fractalization ($\Theta = 0.7$ (a), 5 (b) and 50 (c)).

in order to cover the full range of the ablation mechanism seen experimentally: Coulomb explosion, thermal evaporation and explosive boiling. The range of fractality degrees, which are specific for each ablation mechanism, were defined in our previous work. Here, we confirm that the range remains constant regardless of the nature of the target.

In Figure 6.9, we see the space–time representation of the particle current density. The contour plot representation attached to the 3D representation shows that the maxima shift during expansion. This behavior was seen experimentally through ICCD fast camera imagining. As the current maxima characteristic for each ablation mechanism shifts, it defines a unique slope which describes the expansion velocity of each structure. For the particles ejected through Coulomb explosions, which are described by a low degree of fractalization, a steep slope is defined and thus have a high expansion velocity. The particles are mainly visible in the first moments of expansion. For those ejected through the thermal mechanism, we notice a different slope with a longer lifetime and a bigger spatial expansion (characteristics of a reduced expansion velocity). The last structure is formed mainly by nanoparticles and clusters, and it is defined by a high fractalization degree. We notice that the maximum of the particle current holds its value at small distances even for a long expansion time. This is a well-known trait of this complex structure. If we perform a small calculation by using the initial conditions of our experiments and impose them on the fractal analysis, we can estimate the expansion velocities of each plasma structure. We find for the first structure velocities of 18.7 km/s, for the second structure 2.5 km/s, while for the last structure 710 m/s. The results are in line with the values reported in the literature. Thus, the fractal analysis is a robust one and can be implemented on a wide range of plasmas, regardless of the nature of the targets.

6.3. Space- and time-resolved optical investigations of nanosecond laser-produced plasmas on various geological samples

To achieve our aim of finding general laws that might connect the properties of the mineral with those of the laser ablation plasma, we chose to investigate a wide range of mineral samples. We started with the investigation of simple samples, such as calcite, quartz, or graphite, and continued with more complex samples, such as tourmaline, spodumene, fluorite, chromite, chalcopyrite, galena and celestine. All the minerals were polished using 1000 mesh silicon carbide sandpaper, which led to an average roughness of 0.5 μm for all the samples envisioned in this study.

XRD measurements were performed on powders obtained from the mineral samples. The identified crystalline structures are listed in Table 6.3. The XRD analysis confirms the purity of the minerals,

Table 6.3. Minerals confirmed by the XRD database and their indicatives.

Samples	Confirmed minerals from XRD database
Sample 01	PDF 01-071-3699 $CaCO_3$ Calcite
Sample 02	PDF 01-070-7344 SiO_2 Quartz
Sample 03	PDF 01-078-4715 $(Na_{0.64}K_{0.01}Ca_{0.03})$; $(Al_{0.88}Ti_{0.07}Fe_{1.71}Mn_{0.18}Zn_{0.03}Li_{0.11})$; $(Al_{5.67}Fe_{0.28}Mg_{0.05})$ $((Si_{5.76}Al_{0.24})$ $O_{18})$ $(BO_3)_3$ $(OH)_3$ $((OH)_{0.85}$ $F_{0.15})$ Tourmaline
Sample 04	PDF 00-033-0786 $LiAlSi_2O_6$ Spodumene
Sample 05	PDF 01-077-2245 CaF_2 Fluorite
Sample 06	PDF 01-074-7819 $Al_{0.523}Fe_{0.015}Mg_{0.036}Cr_{1.422}Ni_{0.002}Ti_{0.004}$ Chromite
Sample 07	PDF 00-037-0471 $CuFeS_2$ Chalcopyrite
Sample 08	PDF 01-077-0244 PbS Galena
Sample 09	PDF 01-074-2035 $SrSO_4$ Celestine
Sample 10	PDF 00-025-0284 C Graphite

as we identify only signatures from one particular mineral per each spectrum. The purity of the samples will be essential in finding general laws that could potentially connect the minerals to the properties of the LPPs generated on them, as any changes to the sample structure or composition could influence the properties of the LPP.

For the optical emission spectroscopy and ICCD fast camera imaging results, we showcase here examples for only four selected minerals: spodumene (S4), chromite (S6), galena (S8) and celestine (S9). However, the general trends are discussed for all the investigated minerals and presented in the final section of the chapter.

In Figure 6.10, we have represented the optical emission spectra of the laser-induced plasmas on S4, S6, S8 and S9 in the 350–650 nm spectral range. The spectra have been recorded over a 2 μs gate width, depicting 600 μm across the main expansion axis, over the complete length of the plasma. Each spectrum contains atomic and ionic lines from all the composing elements of each investigated sample. For the chromite plasma, we observe signatures from atomic lines of all the composing elements (Cr I: 357 nm, Al I: 394.4 nm,

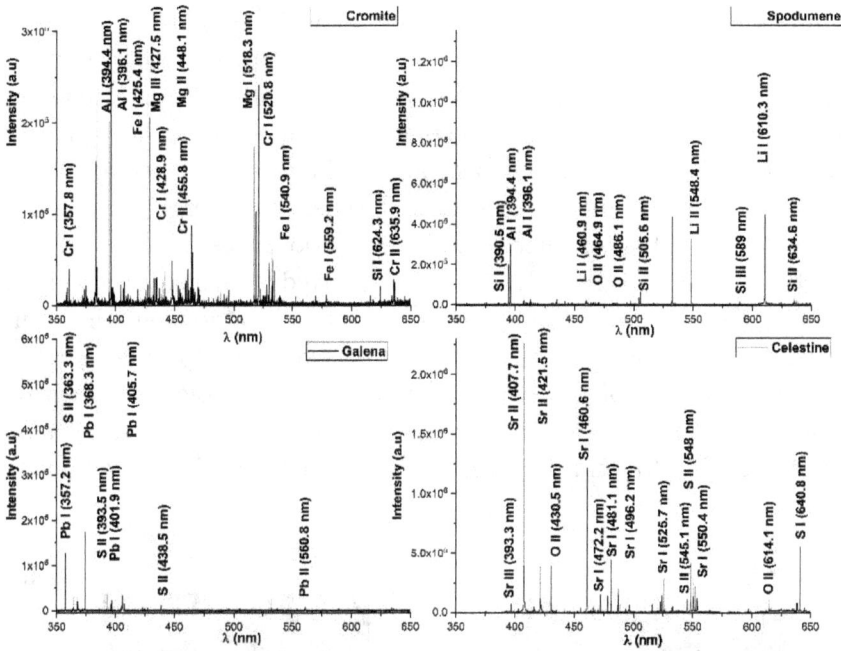

Figure 6.10. Optical emission spectra of LPPs on S6 (a), S4 (b), S8 (c) and S9 (d).

Mg I: 518.3 nm, Fe I: 540.9 nm and Si I: 624.3 nm) and ionic lines for Mg and Cr (Mg II: 448.1 nm, Mg III: 427 nm, Cr II: 455.9 nm) seen throughout the investigated spectral range, with an abundance of Cr atomic lines. For the celestine plasma, we observe signatures for all the atomic (Sr I: 460.6 nm and S: 640 nm) and ionic (Sr III: 393.3 nm, Sr II: 421.5 nm and O II: 430.5 nm) lines. In the spodumene plasma, we observed scattered atomic (Si I: 390.5 nm, Al I: 394.4 nm and Li I: 460.9 nm) and ionic (O II: 464.9 nm, Si II: 505.6 nm, Si III: 589 nm and Li II: 548 nm) lines, with no significant abundance of any specific element. Finally, for the galena plasma, we see an abundance of emission lines for atomic Pb I: 405.7 nm and ionic lines for S II: 438.5 nm and Pb II: 560.8 nm, mainly in the 350–450 nm spectral range. Some differences can be seen in all the collected spectra but also in the four presented in this section. Plasmas generated on samples, such as chromite, presented a bigger abundance in the overall emission lines, which might be a sign of the strong thermalization of the plasma and a lower threshold for excitations of the elemental components of the plasmas. However, there are also samples, such as spodumene or galena, which present more ionized species in the spectra, signaling the presence of species with a lower ionization energy, and thus, the energy is used for ionization processes rather than for excitation ones.

In Figures 6.11(a)–(d), we have plotted the spatial maps at 100 ns after the laser–sample interactions for the aforementioned four plasmas. The spatial maps reveal some exciting aspects regarding the homogeneity of the plasma plumes. We do notice that, within the plasma volume of the investigated chromite plasma, the main elements Cr, Al and Fe are mostly uniformly distributed along

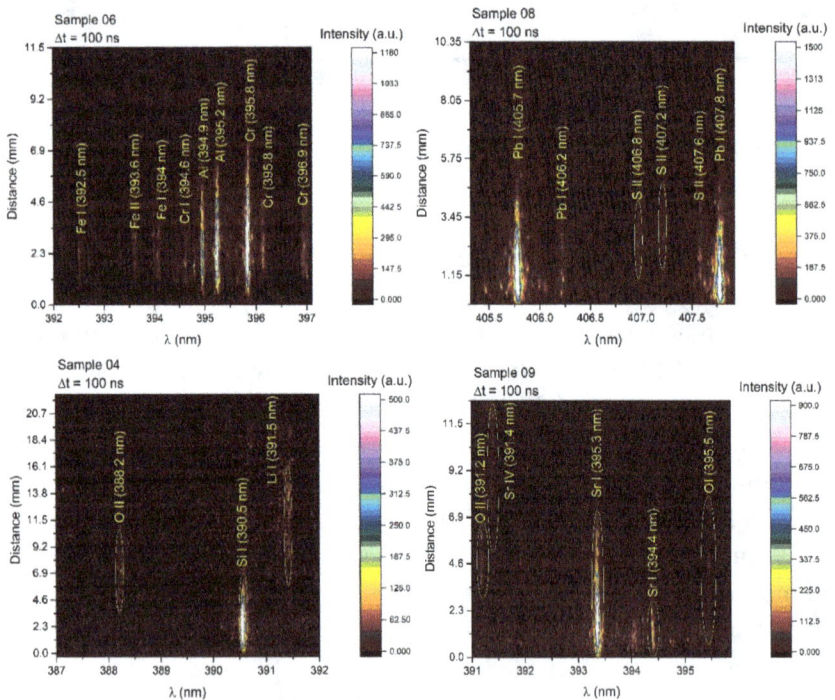

Figure 6.11. Spatial distribution of individual species from spodumene (Sample #4), chromite (Sample #6), galena (Sample #8) and celestine (Sample #9).

the plume length. While for spodumene, galena and celestine, we notice a strong ionic distribution in the proximity of the expansion plume front, as they expand with higher velocities. This result is in good agreement with our previous investigations of complex multi-element plasmas and with the work of Irimiciuc *et al.* (2014, 2018) and Canulescu *et al.* (2009), where they showcased that low-mass elements present higher expansion velocities than "heavier" ones. The individual velocities for each of the observed species are determined by performing a 0.5 nm wide cross-section across the full length of the plume at each moment in time and plotting the spatial displacement of the maximum emission as a function of time. The resulting velocity will characterize the flow of each species. We notice here that the Li I emission in the spodumene sample has a higher expansion velocity compared to that of Si I or even O II. This is a direct result of particle separation with respect to their mass. If we admit that the majority of the neutrals are ejected through a thermal ablation mechanism (Miotello and Kelly, 1995), such as explosive boiling or phase explosion, then lighter elements are expected to have higher velocities due to their lower masses.

The particle distribution within the plasma volume is an important aspect that relates directly to the development of the LIBS technique for understanding the kinetics of the ejected particles and how the distribution within the plasma volume reflects on the overall plasma emission. To assess that aspect of the LPP, we implemented the ICCD fast camera imaging technique on all 10 samples. In Figure 6.12, we have represented snapshots of the LPPs for the selected four samples, covering the expansion over a time span of 2 μs. We observed that the plasmas generated in galena and chromite have a relatively uniform distribution, while the well-known (Semerok *et al.*, 2017; Irimiciuc *et al.*, 2020; Ursu *et al.*, 2010; Harilal *et al.*, 2002; Ojeda *et al.*, 2017) plasma structuring is mainly seen in celestine and spodumene, which also have stronger emissions. This is in good agreement with the results presented in Figure 6.12, where we observe a stronger separation between the ionic and atomic species. The plume-splitting scenarios have been thoroughly investigated by our group (Irimiciuc *et al.*, 2020; Ursu *et al.*, 2010) and confirmed by

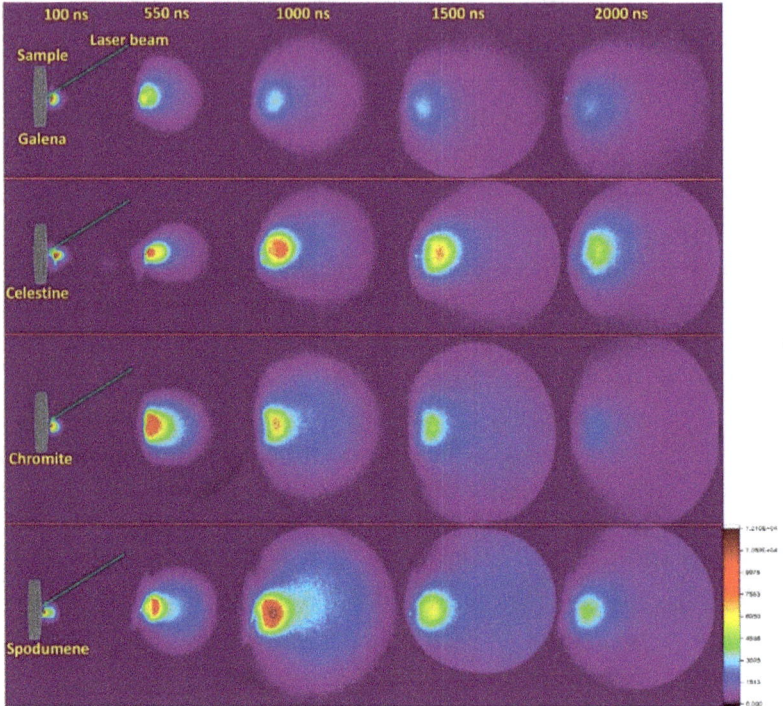

Figure 6.12. ICCD snapshots of LPPs generated on galena, celestine, chromite and spodumene.

other studies (Harilal *et al.*, 2002; Ojeda *et al.*, 2017). The presence of multiple ionized species with different velocities supports the idea of a Coulomb explosion mechanism, while the presence of a strong atomic emission scattered throughout the plasma volume (see the spodumene case) supports the presence of a thermal ejection mechanism as well.

In order to obtain some quantitative information regarding the LPPs on the investigated mineral samples, we also determined the global expansion velocities of each generated plasma, complementing the data characterizing the expansion velocities for each individual element. Similar to the previous approach, this was achieved by performing a 3 nm width cross-section across the length of the plasma plume on the main expansion axis on the ICCD snapshots. The main expansion axis is defined as an orthogonal axis with respect to the target surface, centered at the impact point of the laser beam. The cross-section revealed two emission maxima, each characterizing a different plasma structure, confirming the two-mechanism ablation scenario. Similar results have been reported by our group (Irimiciuc *et al.*, 2014, 2018, 2019, 2020) as well as other groups through both optical (Harilal *et al.*, 2002) and electrical measurements (Thestrup *et al.*, 2002). A comparative study is shown in Figure 6.13, where we represent the expansion velocities for individual species, such as Al, Cr or Ca. We observe that the same species present different velocities

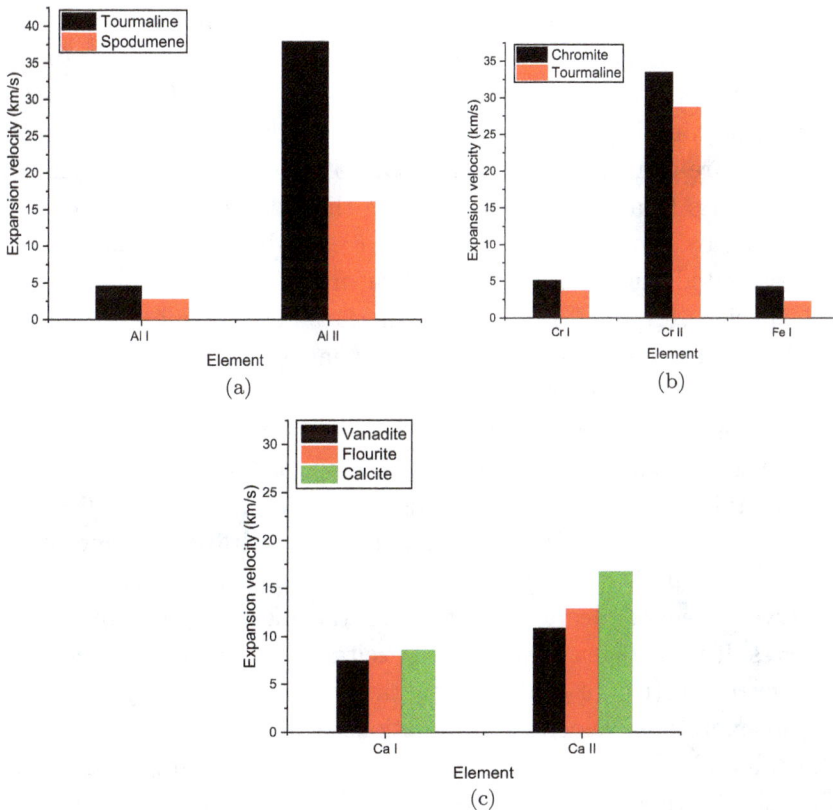

Figure 6.13. Comparison of the velocities of individual species, as seen in the LPPs generated in different minerals.

when ejected from different mineral samples. This can be seen as a fingerprint of the overall physical properties of the samples. For instance, if we consider the spodumene and the tourmaline minerals, the latter contains an amalgam of elements with a high possibility of vacancies in its structure (Dai *et al.*, 2006). The presence of vacancies could lead to a lower melting temperature relating to the higher kinetic energy of the ejected particles.

Furthermore, if we want to analyze the thermal energy of the plasma, we have to estimate the excitation temperature of the plasma. Based on the work by Cristoforetti and Fujimoto (Fujimoto, 1990; Cristoforetti *et al.*, 2013), we estimated the LTE limit for our investigated expansion conditions, which is around 10^{15} cm^{-3}, three orders of magnitude below the values estimated by us, which are in the 10^{16}–10^{17} cm^{-3} range. The LTE limit is in good agreement with other reports for similar conditions in terms of laser fluence and background pressure. Under LTE conditions, the excitation temperature can be used as a measure of the thermal movement of the particles (mainly collisions). Using the Boltzmann plot method, presented by Burger and Herman (2016) and Irimiciuc *et al.* (2020), we estimated the excitation temperature for each of the individual species (Figure 6.14(a)) seen in the emission spectra of each LPP. The choice of calculating individual excitation temperatures is to observe if the heterogeneity in the kinetic energies of the ejected particles is also reflected in the thermal movement. Also, we estimated the global excitation temperature (Figure 6.14(b)) by including all of the excited states recorded. In Figure 6.14(a), we have represented the evolution of the excitation temperature as a function of the melting point of each mineral sample. We observe a general decreasing trend as the melting point increases. The result is also seen in the global excitation temperature presented in Figure 6.14(b). In that representation, we have also plotted the global expansion velocities as functions of the same parameter for the minerals. We notice that, as the excitation temperature and thus the thermal energy of the plasma decrease, the kinetic energy increases. With an increase in the melting point, the thermal ablation mechanism, under identical irradiation conditions, will be reduced, thus tipping the balance toward accelerating the ejected particles

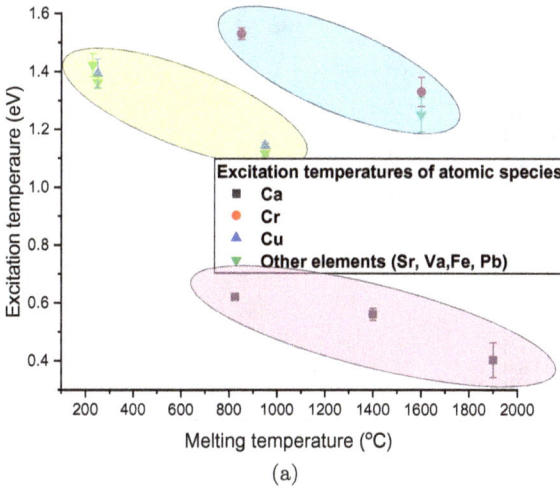

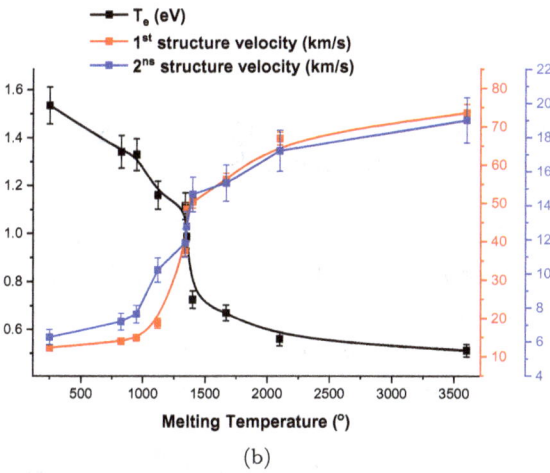

Figure 6.14. Excitation temperatures of the atoms' dependence on the melting point of the mineral samples (a) and the expansion velocity and overall excitation temperature dependence on the melting point (b).

rather than toward breaking the cohesion bonds and evaporating a higher concentration of particles. This representation offers a first empirical law that connects the melting point of the mineral with two important parameters of the plasma: T_{ex} and expansion velocity and shows that we can control the balance between the kinetic and thermal energies of the LPPs.

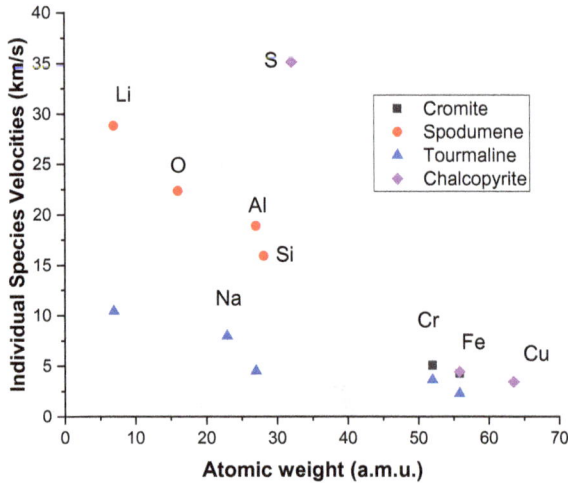

Figure 6.15. Individual species expansion velocities against the atomic masses of different minerals.

Finally, let us compare the individual expansion velocities of the elements composing the minerals as functions of their masses (Figure 6.15). This law has been reported by Canulescu *et al.* (2009), where a series of metals have been investigated and confirmed by our group (Irimiciuc *et al.*, 2020) for femtosecond and picosecond laser ablations. We observe that although we cannot define a single slope for each mineral, the expansion velocities of the individual species define specific negative slopes.

Fractal Analysis:

In the context of our previous experimental data, the ablation plasma dynamics are described through the multifractal theory of motion in the form of hydrodynamic regimes at various resolution scales (multifractal hydrodynamic model, Chapter 3).

Thus, introducing the nondimensional variables,

$$\frac{x}{V_0 \tau_0} = \xi, \quad \frac{t}{\tau_0} = \eta, \tag{6.15}$$

and the nondimensional parameters,

$$\frac{\sigma \tau_0}{\alpha^2} = \mu, \quad \frac{\alpha}{V_0 \tau_0} = \phi, \tag{6.16}$$

with τ_0, the specific time, and $\sigma = \lambda(dt)^{\left[\frac{2}{f(\alpha)}\right]-1}$, the multifractal degree, the normalized forms of the velocities become

$$V_D(\mu, \xi, \eta) = \frac{V_D(x, t)}{V_0} = \frac{1 + \mu^2 \xi \eta}{1 + \mu^2 \eta^2}, \tag{6.17}$$

$$V_F(\mu, \xi, \eta) = \frac{V_F(x, t)}{V_0} = \frac{\mu(\xi - \eta)}{1 + \mu^2 \eta}, \tag{6.18}$$

$$\rho(\mu, \xi, \eta) = \pi^{1/2} \alpha \rho(x, t)$$
$$= \left(1 + \mu^2 \eta^2\right)^{-1/2} \exp\left[-\frac{(\xi - \eta)^2}{\phi^2 \left(1 + \mu^2 \eta^2\right)}\right]. \tag{6.19}$$

From (6.17) and (6.19), the nondimensional non-differentiable current becomes

$$j_D(\mu, \xi, \eta) = \rho(\mu, \xi, \eta) V_D(\mu, \xi, \eta) \tag{6.20}$$
$$= \frac{1 + \mu^2 \xi \eta}{(1 + \mu^2 \eta^2)^{3/2}} \exp\left[-\frac{(\xi - \eta)^2}{\phi^2 \left(1 + \mu^2 \eta^2\right)}\right].$$

Through (6.18) and (6.19), the nondimensional non-differentiable density current is

$$j_F(\mu, \xi, \eta) = \rho(\mu, \xi, \eta) V_F(\mu, \xi, \eta)$$
$$= \mu \frac{(\xi - \eta)^2}{(1 + \mu^2 \eta^2)^{3/2}} \exp\left[-\frac{(\xi - \eta)^2}{\phi^2 \left(1 + \mu^2 \eta^2\right)}\right]. \tag{6.21}$$

Referring to Agop and Paun (2017) and (6.19), the nondimensional specific multifractal potential is

$$Q(\mu, \xi, \eta) = \frac{2Q(x, t)}{V_0^2} = -\frac{\mu^2(\xi - \eta)^2}{2\left(1 + \mu^2 \eta^2\right)^2} - \frac{\mu^2}{(1 + \mu^2 \eta^2)}. \tag{6.22}$$

Let us now calibrate the mathematical model on the empirical data presented in the previous section. According to Agop *et al.* (2020), (6.17) is identified with the expansion velocity of the fast plasma structure (Coulomb structure, V_C), while (6.18) with the slow plasma structure (thermal structure, V_T), and finally, (6.22) can be associated with the electron excitation temperature. Both multifractality through stochasization and the usual time–temperature correlation specific to statistics models (Landau and Lifshitz, 1971) imply the identification of a nondimensional time through the inverse of the nondimensional temperature, i.e. $\eta \equiv T^{-1}$. According to Chapter 3, it results in the specific multifractal potential being defined up to a non-null arbitrary constant. It results thus that the relations (6.17), (6.18) and (6.22) become

$$V_C(\mu, \xi, \eta) = \frac{T\left(T + \mu^2 \xi\right)}{T^2 + \mu^2}, \tag{6.23}$$

$$V_T(\mu, \xi, \eta) = \frac{\mu T (T\xi - 1)}{T^2 + \mu^2}, \tag{6.24}$$

$$T_e(\mu, \xi, \eta) = a - \frac{\mu^2 T^2 (T\xi - 1)^2}{2\left(T^2 + \mu^2\right)^2} - \frac{\mu^2 T^2}{T^2 + \mu^2}, \tag{6.25}$$

$$a = \text{const.}$$

We present in Figure 6.16(a), (c) and (e), the dependences depicted by Equations (6.23)–(6.25). By using an adequate choice of the constants used in the nondimensionalization of both the variables and parameters and identifying T with the melting point of the target material, we can further discuss the empirical data presented in Figure 6.14(b). The mathematical functions were also used to fit the experimental data (Figure 6.16(b), (d) and (f)). We notice that the multifractal model offers a good fit to the experimental data and can be well adjusted for exploring dependences beyond spatial and temporal evolutions, as it was used by Irimiciuc *et al.* (2017). The nature of the fractal model allows us to cross from the usual spatiotemporal scale to dependences on more complex parameters that define the irradiated target. This property of the model is well reflected in the quality of the fit.

The transition from differentiable to non-differentiable dynamics allows for a specific statistic when the correlation between the target properties and the behavior of the ejected particles becomes functional (target–plasma correlation). Furthermore, we admit that the intrinsic properties of the target sample are given implicitly through the multifractality degree $\sigma = \lambda(dt)^{\left[\frac{2}{f(\alpha)}\right]-1}$ and the Gaussian parameter α. Since the dynamics at the two resolution scales

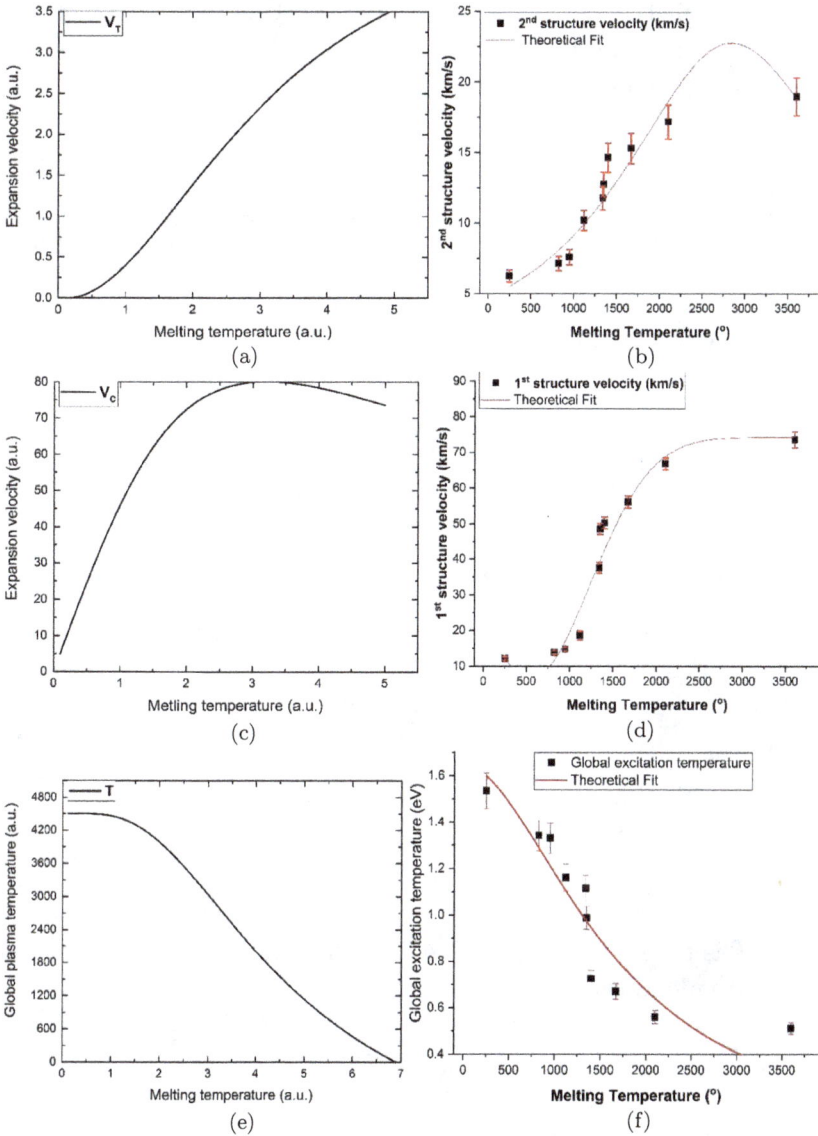

Figure 6.16. Theoretical dependences of plasma temperature (e) and global expansion velocities (a and c) and the theoretical fit of the empirical data (b, d and f).

are hidden in the velocity's expression (see Chapter 3), the critical points of these functions are given by the following restrictions that are respected simultaneously:

$$\frac{\partial V_D(x,t,\sigma,\alpha)}{\partial \sigma} = 0, \quad \frac{\partial V_D(x,t,\sigma,\alpha)}{\partial \alpha} = 0, \qquad (6.26)$$

$$\frac{\partial V_F(x,t,\sigma,\alpha)}{\partial \sigma} = 0, \quad \frac{\partial V_F(x,t,\sigma,\alpha)}{\partial \alpha} = 0. \qquad (6.27)$$

These restrictions imply at a differentiable resolution scale the uniform-type movement:

$$x = V_0 t, \qquad (6.28)$$

while at a non-differentiable resolution scale, they imply the following dependence:

$$\sigma t = \lambda(dt)^{\left[\frac{2}{f(\alpha)}\right]-1} t = \alpha^2 \leftrightarrow \mu\eta = 1. \qquad (6.29)$$

With the constraint (6.29), the differentiable velocity (6.17) takes the form

$$V_D = \frac{1}{2}(1 + \mu\xi) = \frac{1}{2}(1 + \mu\eta). \qquad (6.30)$$

Admitting the functionality presented above and the relations

$$V = \left(\frac{T}{M}\right)^{1/2}, \quad V = \frac{\mu}{V^2}, \qquad (6.31)$$

where T is the nondimensional temperature and M is the nondimensional atomic Mass, (6.30) becomes

$$V_D = \frac{1}{2}\left(1 + \frac{\Omega}{M}\right). \qquad (6.32)$$

We have represented in Figure 6.17(a), the theoretical dependence of (6.32) and the fitting of the empirical data in Figure 6.17(b). In this context, by an adequate choice of the nondimensional variables and an adequate interpretation of V_D, Ω and M, this dependence can describe the general empirical behavior presented in the previous section in Figure 6.17.

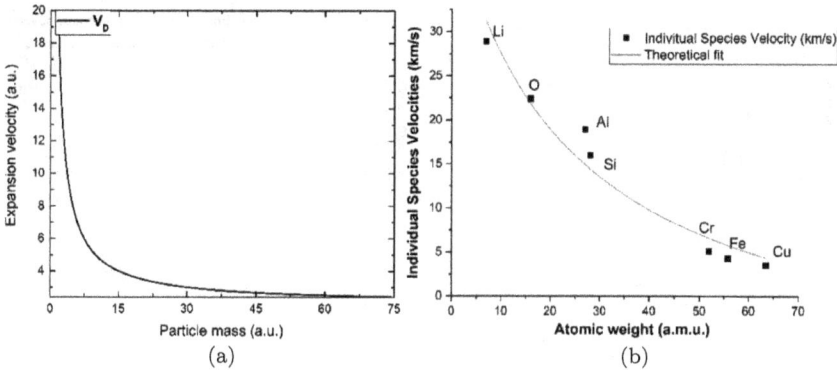

Figure 6.17. Theoretical dependences of individual expansion velocity of the ejected species on the atomic mass (a) and the theoretical fit of the empirical data (b).

Furthermore, with the restriction (6.29), the non-differentiable velocity (6.18) takes the following form:

$$V_F = \frac{1}{2}(\xi\mu - 1) = \frac{1}{2}(\mu\eta - 1). \qquad (6.33)$$

Admitting the following relations:

$$V_F = \left(\frac{\bar{T}}{\bar{M}}\right)^{1/2}, \quad \bar{\Omega} = \frac{\bar{M}}{4}, \qquad (6.34)$$

(6.33) becomes

$$\bar{T} = \bar{\Omega}\left(\frac{\mu}{T} - 1\right)^2. \qquad (6.35)$$

In Figure 6.18, we have represented the relation (6.35). In this context, by adequately choosing the nondimensionalization variable and through a rightful interpretation of $\bar{T}/\bar{\Omega}$, this dependence will now describe the general law presented in Figure 6.14(a). We would also like to mention that the multifractal theoretical model with the appropriate calibration on the experimental investigation techniques can echo the general law presented in the previous section. However, taking into account the previously reported experimental data on pure metals or complex targets, we also notice that our model can also predict the mass dependence of the velocity (Thestrup

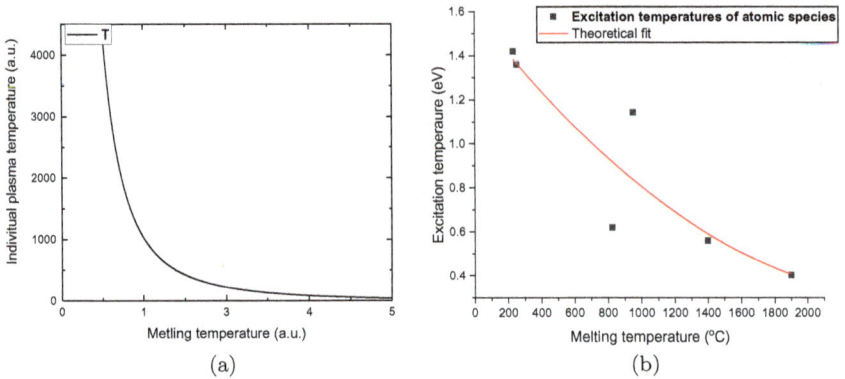

Figure 6.18. Theoretical modeling of the individual species excitation temperature dependences (a) on the melting temperature of the mineral and the fitting of the empirical data (b).

et al., 2002) or the evaporation temperature relation to the overall ablation density (Salle *et al.*, 1999). The empirical data is also fitted (Figure 6.18(b)), where we can see that the majority of experimental points follow the law imposed by the model. Similar empirical data reported by Donnelly *et al.* (2010) follow the same trend as the one predicted by our model, thus we can conclude that the theoretical functions presented have a universal appeal regardless of the experimental technique implemented.

References

Agop M. and Paun V. P. 2017. *On the New Perspectives of Fractal Theory, Fundaments and Applications.* Bucharest: Romanian Academy Publishing House.

Agop M. and Merches I. 2019. *Operational Procedures Describing Physical Systems.* Florida: CRC Press.

Agop M., Mihaila I., Nedeff F., and Irimiciuc S. A. 2020. Charged particle fluctuations in transient plasmas generated by nanosecond laser ablation on Mg, target. *Symmetry Basel*, 12, 292.

Aguilera J. A. and Aragón C. 2007. Multi-element Saha-Boltzmann and Boltzmann plots in laser-induced plasmas. *Spectrochim. Acta-Part B At. Spectrosc.*, 62, 378–385.

Alvey D. C., Morton K., Harmon R. S., Gottfried J. L., Remus J. J., Collins L. M., and Wise M. A. 2010. Laser-induced breakdown spectroscopy-based geochemical fingerprinting for the rapid analysis and discrimination of minerals: The example of garnet. *Appl. Opt.*, 49(13), C168–C180.

Amireh B. S. 1991. Mineral composition of the Cambrian-Cretaceous Nubian Series of Jordan: Provenance, tectonic setting and climatological implications. *Sediment. Geol.*, 71(1–2), 99–119.

Amoruso S., Unitá I., Fisiche S., Federico N., Angelo M. S., Cintia V., Napoli I., Toftmann B., and Schou J. 2004. Thermalization of a UV laser ablation plume in a background gas: From a directed to a di_usionlike flow. *Phys. Rev. E*, 69, 56403.

Amoruso S., Armenante M., Bruzzese R., Spinelli N., Velotta R., and Wang X. 1999. Emission of prompt electrons during excimer laser ablation of aluminum targets. *Appl. Phys. Lett.*, 75, 7–9.

Barker M. and Rayens W. 2003. Partial least squares for discrimination. *J. Chemometrics*, 17(3), 166–173.

Barnolas A., Chiron J. C., and Guérangé B. 1996. Synthèse Géologiqueet Géophysique Des Pyrénées. Volume 1, Introduction, Géophysique, Cycle Hercynien. Bureau de recherches géologiques et minières.

Boyle D. R., Turner R. J. W., and Hall G. E. M. 1998. Anomalous arsenic concentrations in groundwaters of an Island Community, Bowen Island, British Columbia. *Environ. Geochem. Health*, 20(4), 199–212.

Burger M. and Herman J. 2016. Stark broadening measurements in plasmas produced by laser ablation of hydrogen containing compounds. *J. Spectrochim. Acta B*, 122, 118–126.

Bulajic D., Corsi M., Cristoforetti G., Legnaioli S., Palleschi V., Salvetti A., and Tognoni E. 2002. A procedure for correcting self-absorption in calibration free-laser induced breakdown spectroscopy. *Spectrochim. Acta-Part B: At. Spectrosc.*, 57, 339–353.

Canulescu S., Papadopoulou E. L., Anglos D., Lippert T., Schneider C. W., and Wokaun A. 2009. Mechanisms of the laser plume expansion during the ablation of LiMn2O4. *J. Appl. Phys.*, 105, 063107.

Chen Z. and Bogaerts A. 2005. Laser ablation of Cu and plume expansion into 1 atm ambient gas. *J. Appl. Phys.*, 97, 063305.

Colón C. and Alonso-Medina A. 2006. Application of a laser produced plasma: Experimental stark widths of single ionized lead lines. *Spectrochim. Acta B*, 61, 856–863.

Cristoforetti G., Tognoni E., and Gizzi L. A. 2013. Thermodynamic equilibrium states in laser induced plasmas: From the general case to laser-induced breakdown spectroscopy plasmas. *Spectrochim. Acta B*, 90, 1–22.

Dai J. H., Song Y., Xia L., and Wang W. G. 2006. Interactions between carbon species and β-spodumene by first principle calculations. *RSC Adv.*, 7, 70284.

Davari S. A., Hu S., and Mukherjee D. 2016. Calibration-free quantitative analysis of elemental ratios in intermetallic nanoalloys and nanocomposites using laser induced breakdown spectroscopy (LIBS). *Talanta*.

Donnelly T., Lunney J. G., Amoruso S., Bruzzese R., Wang X., and Ni X. 2010. Dynamics of the plasma produced by ultrafast laser ablation of metals. *J. Appl. Phys.*, 108, 043309.

Dunham R. J. 1962. Classification of carbonate rocks according to depositional textures. 38, 108–121.

Dunham R. J. 1962. Classification of carbonate rocks according to depositional textures. In *Classification of Carbonate Rocks — A Symposium*, W. E. Ham (Ed.) (pp. 108–121). Vancuver, BC, Canada: American Association of Petroleum Geologists.

Dyar M. D., Carmosino M. L., Breves E. A., Ozanne M. V., Clegg S. M., and Wiens R. C. 2012. Comparison of partial least squares and lasso regression techniques as applied to laser-induced breakdown spectroscopy of geological samples. *Spectrochimica Acta — Part B At. Spectrosc.*, 70, 51–67.

El Haddad J. Canioni L., and Bousquet B. 2014. Good practices in LIBS analysis: Review and advices. *Spectrochim. Acta-Part B At. Spectrosc.*, 101, 171–182.

Enescu F., Irimiciuc S. A., Cimpoesu N., Bedelean H., Bulai G., Gurlui S., and Agop M. 2019. Investigations of laser produced plasmas generated by laser ablation on geomaterials. *Exp. Theor. Aspects Symmet. Basel*, 11, 1391.

Eppler A. S., Cremers D. A., Hickmott D. D., Ferris M. J., and Koskelo A. C. 1996. Matrix effects in the detection of Pb and Ba in soils using laser-induced breakdown spectroscopy. *Appl. Spectrosc.*, 50(9), 1175–1181.

Ershov-Pavlov E. A., Katsalap K. Y., Stepanov K. L., and Stankevich Y. A. 2008. Time-space distribution of laser-induced plasma parameters and its influence on emission spectra of the laser plumes. *Spectrochim. Acta Part B: At. Spectrosc.*, 63, 1024–1037.

Friedman R. M., Monger J. W. H., and Tipper H. W. 1990. Age of the Bowen Island Group, Southwestern Coast Mountains, British Columbia. *Can. J. Earth Sci.*, 27(11), 1456–1461.

Fujimoto T. 1990. Validity criteria for local thermodynamic equilibrium in plasma spectroscopy. *Phys. Rev. A*, 42, 6588–6601.

González R., Lucena P., Tobaria L. M., and Laserna J. J. 2009. Standoff LIBS detection of explosive residues behind a barrier. *J. Anal. At. Spectrom.*, 24, 1123–1126.

Haddad J. E., Canioni L., and Bousquet B. 2014. Good practices in LIBS analysis: Review and advices. *Spectrochim. Acta — Part B: At. Spectrosc.*, 101, 171–182.

Hahn D. W. and Omenetto N. 2010. Laser-induced breakdown spectroscopy (LIBS), Part I: Review of basic diagnostics and plasma particle interactions: Still-challenging issues within the analytical plasma community. *Appl. Spectrosc.*, 64(12), 335–366.

Hai R., Mao X., Chan G. C.-Y., Russo R. E., Ding H., and Zorba V. 2018. Internal mixing dynamics of Cu/Sn-Pb plasmas produced by femtosecond laser ablation. *Spectrochim. Acta B*, 148, 92–98

Harilal S. S., Bindhu C. V., Tillack M. S., Najmabadi F., and Gaeris A. C. 2002. Plume splitting and sharpening in laser-produced aluminium plasma. *J. Phys. D. Appl. Phys.*, 35, 2935–2938.

Harilal S. S., Bindhu C. V., Tillack M. S., Najmabadi F., and Gaeris A. C. 2003. Internal structure and expansion dynamics of laser ablation plumes into ambient gases. *J. Appl. Phys.*, 93, 2380–2388.

Harmon R. S., Russo R. E., and Hark R. R. 2013. Applications of laser-induced breakdown spectroscopy for geochemical and environmental analysis: A comprehensive review. *Spectrochim, Acta B*, 87, 1–26.

Hermann J., Axente E., Pelascini F., and Craciun V. 2019. Analysis of multi-elemental thin films via calibration-free laser-induced breakdown spectroscopy. *Anal. Chem.*, 91, 2544–2550.

Hervé K. S., Jain J., Bol'Shakov A., Lopano C., McIntyre D., and Russo R. 2016. Determination of elemental composition of Shale Rocks by laser induced breakdown spectroscopy. *Spectrochim. Acta — Part B: At. Spectrosc.*, 122, 9–14.

Huang R., Yu Q., Tong Q., Hanga W., He J., and Huang B. 2009. Influence of wavelength, irradiance, and the buffer gas pressure on high irradiance laser ablation and ionization source coupled with an orthogonal time of flight mass spectrometer. *Spectrochim. Acta B*, 64, 255–261.

Irimiciuc S. A., Boidin R., Bulai G., Gurlui S., Nemec P., Nazabal V., and Focsa C. 2017a. Laser ablation of (GeSe2)100-x (Sb2Se3)x chalcogenide glasses: Influence of the target composition on the plasma plume dynamics. *Appl. Surf. Sci.*, 418, 594–600.

Irimiciuc S. A., Gurlui S., Nica P., Focsa C., and Agop M. 2017b. A compact non-differential approach for modeling laser ablation plasma dynamics. *J. Appl. Phys.*, 121, 83301.

Irimiciuc S. A., Bulai G., Agop M., and Gurlui S. 2018a. Influence of laser-produced plasma parameters on the deposition process: In situ space- and time-resolved optical emission spectroscopy and fractal modeling approach. *Appl. Phys A: Mater.*, 124, 615.

Irimiciuc S. A., Mihaila I., and Agop M. 2014. Experimental and theoretical aspects of a laser produced plasma. *Phys. Plasmas*, 21, 93509.

Irimiciuc S. A., Bulai G., Gurlui S., and Agop M. 2018b. On the separation of particle flow during pulse laser deposition of heterogeneous materials — A multi-fractal approach. *Powder Tech.*, 339, 273–280.

Irimiciuc S. A., Bulai G., Agop M., and Gurlui S. 2018c. Influence of laser-produced plasma parameters on the deposition process: In situ space- and time-resolved optical emission spectroscopy and fractal modeling approach. *Appl. Phys. A: Mater. Sci. Process.*, 124, 615.

Irimiciuc S. A., Gurlui S., and Agop M. 2019. Particle distribution in transient plasmas generated by ns-laser ablation on ternary metallic alloys. *Appl. Phys. B*, 125, 190.

Irimiciuc S. A., Nica P. E., Agop M., and Focsa C. 2020. Target properties — plasma dynamics relationship in laser ablation of metals: Common trends for fs, ps and ns irradiation regimes. *Appl. Surf. Sci.*, 506, 144926.

Kramida A. and Ralchenko Y. 2014. NIST ASD Team, NIST Atomic Spectra Database Lines Form, NIST At. Spectra Database (Ver. 5.2).

Landau L. D. and Lifshitz E. M. 1971. *Theoretical Physics*, Vol. 4. Pergamon.

Leitch, H. C. B. 1947. Contributions to the Geology of Bowen Island.

Makhlouf I. M. and Abed A. M. 1991. Depositional facies and environments in the Umm Ishrin sandstone formation, Dead Sea Area, Jordan. *Sediment. Geol.*, 71(3–4), 177–187.

Matroodi F. and Tavassoli S. H. 2014. Simultaneous Raman and laser-induced breakdown spectroscopy by a single setup. *Appl. Phs. B*, 117(4), 1081–1089.

Merches I. and Agop M. 2016. *Differentiability and Fractality in Dynamics of Physical Systems*. New Jersey: World Scientific.

Migoń P. and Goudie A. 2014. Sandstone geomorphology of South-West Jordan, Middle East. *Quaestiones Geographicae*, 33(3), 123–130.

Miotello A. and Kelly R. 1995. Critical assessment of thermal models for laser sputtering at high fluences. *Appl. Phys. Lett.*, 67, 3535.

Ojeda-G-P A., Schneider C. W., Döbeli M., Lippert T., and Wokaun A. 2017. Plasma plume dynamics, rebound, and recoating of the ablation target in pulsed laser deposition. *J. Appl. Phys.*, 121135306.

O'Mahony D., Lunney J., Dumont T., Canulescu S., Lippert T., and Wokaun A. 2007. Laser-produced plasma ion characteristics in laser ablation of lithium manganate. *Appl. Surf. Sci.*, 254, 811–815.

Popov A. M., Zaytsev S. M., Seliverstova I. V., Zakuskin A. S., and Labutin T. A. 2018. Matrix effects on laser-induced plasma parameters for soils and ores. *Spectrochim. Acta-Part B: At. Spectrosc.*, 148, 205–210.

Puretzky A. A., Geohegan D. B., Haufler R. E., Hettich R. L., Zheng X. Y., and Compton R. N. 1993. Laser ablation of graphite in different buffer gases. *AIP Conf. Proc.*, 288, 365–374.

Rao K. H., Smijesh N., Nivas J. J., and Philip R. 2016. Ultrafast laser produced zinc plasma: Stark broadening of emission lines in nitrogen ambient. *Phys. Plasmas*, 23, 43503.

Russell H. S., Russo R. E., and Hark R. R. 2013. Applications of laser-induced breakdown spectroscopy for geochemical and environmental analysis: A comprehensive review. *Spectrochim. Acta — Part B: At. Spectrosc.*, 87, 11–26.

Rodríguez L., Cuevas J., and Tubía, J. M. 2014. Structural evolution of the sierras interiores (Aragón and Tena Valleys, South Pyrenean Zone): Tectonic implications. *J. Geol.*, 122, 99–111

Roigé M., Gómez-Gras D., Remacha E., Daza R., and Boya S. 2016. Tectonic control on sediment sources in the Jacabasin (Middle and Upper Eocene of the South-Central Pyrenees). *C. R. Geosci.*, 348, 236–245.

Salle B., Gobert O., Meynadier P., Perdrix M., Petite G., and Semerok A. 1999. Femtosecond and picosecond laser microablation: Ablation efficiency and laser microplasma expansion. *Appl. Phys. A*, 69, 381–383.

Salle B., Chaleard C., Detalle V., Lacour J.-L., Mauchien P., Nouvellon C., and Semerok A. 1999. Laser ablation efficiency of metal samples with UV laser nanosecond pulses. *Appl. Surf. Sci.*, 138–139, 302–305.

Sanghapi H. K., Jain J., Bol'Shakov A., Lopano C., McIntyre D., and Russo R. 2016. Determination of elemental composition of shale rocks by laser induced breakdown spectroscopy. *Spectrochim. Acta-Part B: At. Spectrosc.*, 122, 9–14.

Semerok A., Fomichev S. V., Jabbour C., Lacour J.-L., Tabarant M., and Chartier F. 2017. Multi-parametric modeling of solid sample heating by nanosecond laser pulses in application for nano-ablation. *Spectrochim. Acta B*, 136, 51–55.

Senesi G. S. 2014. Laser-induced breakdown spectroscopy (LIBS) applied to terrestrial and extraterrestrial analogue geomaterials with emphasis to minerals and rocks. *Earth Sci. Rev.*, 139, 231–267.

Shaikh N. M., Hafeez S., and Baig M. A. 2007. Comparison of zinc and cadmium plasma parameters produced by laser-ablation. *Spectrochim. Acta B*, 62, 1311–1320.

Sharma S. K., Misra A. K. P. G., Lucey R. C., and Clegg Wiens S. M. 2007. Combined remote LIBS and Raman spectroscopy at 8.6 m of sulfur-containing minerals, and minerals coated with hematite or covered with basaltic dust. *Spectrochim. Acta A-M*, 68(4), 1036–1045.

Sibuet J. C., Srivastava, S. P., and Spakman W. 2004. Pyrenean orogeny and plate kinematics. *J. Geophys. Res.*, 109, B08104.

Thestrup B., Toftmann B., Schou J., Doggett B., and ad Lunney J. G. 2002. Ion dynamics in laser ablation plumes from selected metals at 355 nm. *Appl. Surf. Sci.*, 197–198, 175–180.

Tognoni E., Cristoforetti G., Legnaioli S., Palleschi V., Salvetti A., Müller M., and Gornushkin I. 2007. A numerical study of expected accuracy and precision in calibration-free laser-induced breakdown spectroscopy in the assumption of ideal analytical plasma. *Spectrochim. Acta-Part B: At. Spectrosc.*, 62, 1287–1302.

Ursu C., Pompilian O. G., Gurlui S., Nica P., Agop M., Dudeck M., and Focsa C. 2010. Al2O3 ceramics under high-fluence irradiation: Plasma plume dynamics through space and time-resolved optical emission spectroscopy. *Appl. Phys A: Mater.*, 101, 153–159.

Valero Garcés B. L. and Aguilar J. G. 1992. Shallow carbonate lacustrine facies models in the Permian of the Aragon-Bearn Basin (Western Spanish-French Pyrenees). *Carbonates Evaporites*, 7(2), 94–107.

Waltham T. 1994. The sandstone fantasy of Petra. *Geol. Today*, 10, 105–111.

Waltham T. 1994. The sandstone fantasy of petra. *Geology Today*, 10(3), 105–111.

Xie S., Xu T., Han X., Lin Q., and Duan Y. 2017. Accuracy improvement of quantitative LIBS analysis using wavelet threshold de-noising. 32(3), 629–637.

Xu T., Liu J., Shi Q., He Y., Niu G., and Duan Y. 2016. Multi-elemental surface mapping and analysis of carbonaceous Shale by laser-induced breakdown spectroscopy. *Spectrochim. Acta B: At. Spectrosc.*, 115(January), 31–39.

Conclusions and Perspectives

A general overview on the history of laser ablation is attempted here, with the emphasis falling on the evolution of the diagnostic tools used nowadays with regularity. The advancement of experimental investigations not only aided the development of new techniques but also enhanced the understanding of the fundamental aspects of laser–matter interactions. Several complimentary techniques were extensively presented, aimed at revealing the facets of the same process. The aim of this manuscript was to present the fundamentals of laser ablation and transient plasma dynamics from the perspective of a multifractal paradigm. Over the years, the model managed to reflect well the dynamics reported using both electrical and optical investigation techniques. The specific monofractal dynamics of Nottale's scale relativity theory (NSRT) are expanded to multi-fractal dynamics in the form of the multifractal theory of motion. Consequently, a short presentation of the non-differentiability of the multifractal theory was given. A quick overview of Shannon's information was provided by defining this information and its invariance regarding any transformation group. From a stochastic point of view, if the variables of the transformation group are uniformly distributed, then the principle that minimizes the information is identified with that of the maximum Onicescu's informational energy. The Lagrange multipliers method is used in the minimization of information and the explanation of the information in the case of a complex system exhibiting dynamic radial symmetry is given. The correlation between Shannon's information "subjected" to the "qualities" described in the multifractal theory of motion in its hydrodynamic form implies Newtonian-type behaviors dependent on scale resolution (i.e. interactions of a multifractal Newtonian-type).

The analysis of these behaviors in the framework of the multifractal theory of motion in its hydrodynamic form involves, through a set of complex variables, motion geodesics (i.e. trajectories), with their analytic expressions given by means of conics. In such a context, we showed that the center of the Newtonian-type multifractal force is different from the center of the multifractal trajectory, the measure of this difference being the eccentricity, which depends on the initial conditions. The eccentricities' geometry (the initial conditions' geometry) became, through the Cayley–Klein metrization principle, the Lobachevsky plane geometry. The harmonic mappings between the usual space and the Lobachevsky plane in a Poincaré metric mean that the Ernst potential of general relativity becomes, in essence, of a classical nature. This represents a harmonic application from the usual space to the matter–field space without making use of the concept of material structure, so necessary for the general relativity theory in Einstein's view. Thus, a "physics" of the initial conditions "specified" by Newtonian–type multifractal motions becomes functional, to which it corresponds a Lobachevsky geometry, through the Cayley–Klein metrization principle. Furthermore, the algebraic structure associated with this "physics" is isomorphic with the algebraic structure "dictated" by the Ernst principle upon the gravitational field in vacuum. As such, the gravitational field revealed itself in two separate instances — the multifractal theory of motion (presenting a joint group invariant of $SL(2R)$ type, which can be determined by means of Stoka's procedure). Several types of harmonic mappings, given by both scale resolution and temporal ordering, were graphically explained. The current theoretical model becomes applicable to any case of the existence of a real conic, as it is easy to demonstrate that the Cayleyan metric attached to this conic is a metric generated by the transformation group of $SL(2R)$ type, which leaves it invariant.

In order to calibrate our part of our model, several types of plasma were considered. Reports on investigations of transient plasmas generated by nanosecond laser ablation on a Mg target at high irradiation fluences are given by means of the Langmuir probe method and confronted with the multifractal model. The electrical investigations revealed the presence of a dual structure for the ionic and electronic saturation currents. Quasi-periodic fluctuations were observed for short evolution times and a classical Coulomb–Maxwellian shape for longer times. Two frequencies were computed under all irradiation conditions: the first of a few tens of MHz and the second one of a few MHz. The values for these frequencies did not reflect on the electronic counterpart; therefore, this behavior is not a collective oscillation of the ions but rather a selection of the ionic groups based on their mass/charge ratio. Each frequency characterizes an individual plasma structure and is induced by different double layers, which can accelerate the charged particles. Space- and time-resolved investigations were performed, and an exponential decrease in the main plasma parameters during expansion was observed, coupled with an increase in the plasma potential at high measurement distances. Similar empirical evolutions were found to be described by a steep increase followed by a saturation at higher fluences.

These dynamics of laser-produced plasmas were further described through the fractal theory of motion given by Schrödinger regimes of fractal type. The calibration of such dynamics through a fractal-type tunneling effect for physical systems with spontaneous symmetry breaking allowed the self-structuring of a laser-produced plasma into two substructures based on its separation into different modes, as well as the determination of some characteristics involved in the self-structuring process. The mutual conditionings between the two substructures are given as a joint invariant function of the action of the two isomorphous groups of $SL(2R)$. Their isomorphism implies the substructure self-modulation of the dynamics amplitude through a Stoler-type transformation (i.e. through charge creation — annihilation processes). Other nonlinear behaviors of transient plasma structural units were analyzed. For irrotational motions of the transient plasma structural units, the geodesic equation takes the form of a Schrodinger equation of multifractal type. In the stationary case of this equation, a "hidden symmetry" of $SL(2R)$ type is highlighted, a situation in which various "synchronization modes" among the structural units of a complex system become functional. For a Riccati-type gage, these "synchronization modes" can be seen as nonlinear behaviors in the form of period doubling of the damped fluctuations of quasi-periodicity, intermittencies, etc. In such a manner, possible scenarios toward chaos, without concluding in chaos (nonmanifest chaos), can be mimed. Expanding the reach of the model from laser-produced plasmas to a more general concept of complex fluids, the dynamics of a complex fluid were investigated in the framework of a non-differential model. Separations into multiple structures were observed for the multifractal system containing elements with various physical properties. The formation of complex fluid structures is directly related to the interaction between the specific structural units, and it is represented here by the complex phase of the velocity field and the fractalization of the particle's trajectories. The non-differential system of equations was simplified by analyzing only two main directions. The fluid splits into multiple structures symmetrically on the main expansion axis.

The theoretical model was compared with experimental investigations of transient plasmas generated by laser ablation on a complex (multi-element) metallic target. The evolution of the plasma plume was investigated by means of ICCD fast camera imaging and optical emission spectroscopy (OES). The ICCD fast camera imaging revealed the formation of two plasma structures in the main expansion direction, coupled with a similar phenomenon in the transversal direction. This affects the angle expansion of the plasma, which in turn affects the spatial distribution of the composing element. OES was implemented to investigate the dynamics of the individual species found in the plume. The recorded emission lines were associated with all elements of the target. During expansion, each species presented different electron temperatures and thus different thermal velocities. The heterogeneity of the plasma plume velocity field is in good agreement with the theoretical assumption presented in the framework of the non-differential model.

ICCD imaging revealed the splitting of the laser-produced plasmas into two different structures, expanding with different velocities.

An angular distribution of the front velocity was reconstructed for each of the two plasmas. Space- and time-resolved OES was performed in order to study the individual dynamics of the ejected species. The collected spectra revealed the presence of atomic and ionic species. The OES results correlated to the ones obtained by the ICCD images suggest that the ions are found predominantly in the fast structure, while the atoms are the main components of the slow structure. The spatial distribution of the excitation temperatures determined using the Boltzmann plot revealed strong fluctuations for ions, while a smoother distribution was found for atoms. The species velocities were correlated to the properties of the elements found in the target (mass and conductivity). The SEM images of the deposited samples revealed a more uniform growth of the Al thin film, and the EDX mapping showed Ag segregations on the surface of the film deposited on the alloy target. This result was associated with the cohesion energy and the heat of fusion values of the main elements.

To further extend the reach of our model, we departed from the metallic nature of the targets and looked into geological samples. The approach is natural because, seeing the plasma as a complex fluid, a multi-element heterogenous plasma seems the ideal candidate for testing. Various samples with different geomorphological backgrounds collected from the Northern Hemisphere were investigated. Optical microscopy and XRD analysis were used to identify the minerals present in the rocks, while EDX measurements revealed the composition of the samples.

ICCD fast camera imaging and OES revealed a split of the ejected cloud into three structures. OES allowed the identification of all the elements in the target and confirmed the results obtained through EDX and XRD. Important differences were found between the values of each species and their corresponding ions, a result discussed in the framework of selective heating by the incoming pulse. The values of the global plasma structure and those of the individual species temperatures were connected to the presence of calcite in the samples. Other aspects, such as target porosity, were found to play smaller roles in the plasma plume dynamics. The samples with the highest velocities have low concentrations of calcite, while the sedimentary rocks presented low expansion velocities and high excitation temperatures, which is in line with the effects of a strong thermalized ablation process. The mapping of the emission lines of each of the composing elements allowed us to showcase the differences in the plasma homogeneity based on their kinetic energies. This is also seen through ICCD, where the plume-splitting phenomenon is shown to be present in all 10 samples. The optical investigation revealed an interdependency between the expansion velocity of plasma structures, the excitation temperatures for the individual species and the melting temperatures of the elements composing the minerals. Also, the individual expansion velocities were found to be dependent on the atomic masses of the species.

Throughout this manuscript, we aimed to first present the multifractal NSRT model for laser-produced plasmas in a unique way. This aims to be the first comprehensive depiction of the model as implemented for laser-produced plasmas. The further development of the model needs to focus on the intricate dynamics of charged particles within the framework of the pulsed laser deposition (PLD) process. As PLD becomes closer to an industrial tool, it becomes imperative to have comprehensive models that could attest to the existing phenomena and deliver on the connections between the properties of the target, laser, plasma and those of the deposited film, which is a complex puzzle that needs to be solved.

Index

www.ingramcontent.com/pod-product-compliance
Lightning Source LLC
Chambersburg PA
CBHW050558190326
41458CB00007B/2087